T0344894

GUIDELINES FOR
# MANAGING ABNORMAL SITUATIONS

This book is one in a series of process safety guidelines and concept books published by the Center for Chemical Process Safety (CCPS). Please go to *www.wiley.com/go/ccps* for a full list of titles in this series.

It is sincerely hoped that the information presented in this document will lead to an even more impressive safety record for the entire industry. However, the American Institute of Chemical Engineers, its consultants, the CCPS Technical Steering Committee and Subcommittee members, their employers, their employers' officers and directors, the Abnormal Situation Management® Consortium (ASMC) and its members, and Baker Engineering and Risk Consultants, Inc. (BakerRisk®), and its employees do not warrant or represent, expressly or by implication, the correctness or accuracy of the content of the information presented in this document. As between (1) American Institute of Chemical Engineers, its consultants, CCPS Technical Steering Committee and Subcommittee members, their employers, their employers' officers and directors, the ASMC members, and BakerRisk, and its employees and (2) the user of this document, the user accepts any legal liability or responsibility whatsoever for the consequences of its use or misuse.

# Guidelines for Managing Abnormal Situations

**Center for Chemical Process Safety**

**Of The**

**American Institute of Chemical Engineers**

New York, NY

This edition first published 2023
© 2023 the American Institute of Chemical Engineers

A Joint Publication of the American Institute of Chemical Engineers and John Wiley & Sons, Inc.

*Registered Office*
John Wiley & Sons, Inc., 111 River Street, Hoboken, NJ 07030, USA

For details of our global editorial offices, customer services, and more information about Wiley products visit us at www.wiley.com.

Wiley also publishes its books in a variety of electronic formats and by print-on-demand. Some content that appears in standard print versions of this book may not be available in other formats.

*Library of Congress Cataloging-in-Publication Data Applied for:*
Hardback ISBN: 9781119862871

Cover Images: Dow Chemical Operations, Stade, Germany/
Courtesy of Dow Chemical Company
manyx31/Getty Images;
Creativ Studio Heinemann/Getty Images
SKY10037595_103122

# TABLE OF CONTENTS

# LIST OF FIGURES

# LIST OF TABLES

# LIST OF EXAMPLE INCIDENTS

# ACRONYMS AND ABBREVIATIONS

| | |
|---|---|
| ADIRU | Air Data Inertial Reference Units (Ch. 7) |
| AIChE | American Institute of Chemical Engineers (Preface) |
| AIM | Asset Integrity Management (Ch. 3) |
| AOA | Angle of Attack (Ch. 7) |
| AOPS | Automatic Overfill Protection Systems (Ch. 3) |
| APC | Advanced Process Control (Ch. 3) |
| API | American Petroleum Institute (Ch. 2) |
| ASM® | Abnormal Situation Management® (Ch. 1) |
| ASMC | Abnormal Situation Management® Consortium (Ch. 1) |
| BLEVE | Boiling Liquid Expanding Vapor Explosion (Ch. 3) |
| CAS | Computerized Air Speed (Ch. 7) |
| CCPS | Center for Chemical Process Safety (Preface) |
| CDU | Crude Distillation Unit (Ch. 7) |
| CIMAH | Control of Industrial Major Accident Hazards (Ch. 7) |
| COMAH | Control of Major Accident Hazards (Ch. 7) |
| COO | Conduct of Operations (Ch. 2) |
| CPC | Critical Process Controller (Ch. 7) |
| CSB | Chemical Safety Board (Ch. 3) |
| DCS | Distributed Control Systems (Ch. 2) |
| ECAM | Electronic Centralized Aircraft Monitoring (Ch. 7) |
| EFCS | Electronic Flight Control System (Ch. 7) |
| EHSS | Environmental Health, Safety and Security (Ch. 4) |
| FCCU | Fluidized Catalytic Cracker Unit (Ch. 7) |

| | |
|---|---|
| FCPC | Flight Control Primary Computer [aka PRIM] (Ch. 7) |
| FCSC | Flight Control Secondary Computer [aka SEC] (Ch. 7) |
| FDR | Flight Data Recorder (Ch. 7) |
| FMEA | Failure Modes and Effects Analysis (Ch. 5) |
| GCPS | Global Congress on Process Safety (Ch. 5) |
| GEMS | Generic Error-modelling System (Ch. 3) |
| GPWS | Ground Proximity Warning System (Ch. 7) |
| HAZID | Hazard Identification (Ch. 5) |
| HAZOP | Hazard and Operability Study (Ch. 3) |
| HF | Hydrogen Fluoride (Ch. 3) |
| HIRA | Hazard Identification and Risk Analysis (Ch. 3) |
| HMA | Highly Managed Alarm (Ch. 4) |
| HMI | Human Machine Interface (Ch. 3) |
| HRA | Human Reliability Analysis (Ch. 3) |
| IOGP | International Association of Oil and Gas Producers (Ch. 5) |
| IOW | Integrity Operating Window (Ch. 4) |
| ITCZ | Inter-Tropical Convergence Zone (Ch. 7) |
| ITPM | Inspection, Testing, and Preventive Maintenance (Ch. 6) |
| LCN | Light Cycle Naphtha (Ch. 7) |
| LOPA | Layer of Protection Analysis (Ch. 4) |
| LOPC | Loss of Primary Containment (Ch. 6) |
| LPG | Liquefied Petroleum Gas (Ch. 3) |
| MCAS | Maneuvering Characteristics Augmentation System (Ch7) |
| MIC | Methyl Isocyanate (Ch. 3) |
| MOC | Management of Change (Ch. 3) |
| MOOC | Management of Organizational Change (Ch. 5) |
| ND | Navigation Display (Ch. 7) |
| OD | Operational Discipline (Ch. 2) |
| PF | Pilot (who is) Flying (Ch. 7) |
| PFD | Primary Flight Display (Ch. 7) |
| PHA | Process Hazards Analysis (Ch. 3) |
| PNF | Pilot Not Flying (Ch. 7) |
| PORV | Pilot Operated [Pressure] Relief Valve (Ch. 5) |
| PSID | Process Safety Incident Database (Ch. 3) |
| PSM | Process Safety Management (Ch. 7) |

| PSSR | Pre-Startup Safety Review (Ch. 5) |
|------|-----------------------------------|
| PSV | Pressure Safety Valve (Ch. 6) |
| RAGAGEP | Recognized And Generally Accepted Good Engineering Practice (Ch. 3) |
| RBI | Risk Based Inspection (Ch. 3) |
| RBPS | Risk Based Process Safety (Ch. 1) |
| RCM | Reliability Centered Maintenance (Ch. 3) |
| SA | Situational Awareness (Ch. 3) |
| SIS | Safety Instrumented System (Ch. 3) |
| SME | Subject Matter Expert (Ch. 4) |
| SMS | Safety Management Systems (Ch. 7) |
| SOP | Standard Operating Procedure (Ch. 4) |
| TOH | Transient Operation HAZOP (Ch. 5) |
| UCDS | User Centered Design Services (Ch. 1) |
| VCE | Vapor Cloud Explosion (Ch. 2) |
| VDU | Vacuum Distillation Unit (Ch. 7) |

# GLOSSARY

**Abnormal Situation**

A disturbance in an industrial process with which the basic process control system of the process cannot cope.

Note: In the context of a hazard evaluation, synonymous with deviation.

**Abnormal Situation Management**

Abnormal Situation Management, or Managing Abnormal Situations, refers to a comprehensive process for improving performance which addresses the entire plant population. It promotes effective utilization of all available resources—i.e., hardware, software, and people, including the proactive or reactive intervention activities of members of the operations team, to achieve safe and efficient operations. Abnormal Situation Management is achieved through prevention, early detection, and mitigation of abnormal situations.

**Advanced Process Control**

Advanced process control refers to techniques including multi-variable control, inferential control, feedforward, and decoupling. Multiple single-loop controllers are adjusted in unison, to satisfy constraints and attain optimization objectives while adhering to safe operating limits. Advanced process control techniques often use model-based software to direct the process operation. These applications require that the process model created accurately represents the process dynamics.

**Asset Integrity Management**

A process safety management system for ensuring the integrity of assets throughout their life cycle.

| | |
|---|---|
| **Boiling Liquid Expanding Vapor Explosion (BLEVE)** | A type of rapid phase transition in which a liquid contained above its atmospheric boiling point is rapidly depressurized, causing a nearly instantaneous transition from liquid to vapor with a corresponding energy release. A BLEVE of flammable material is often accompanied by a large aerosol fireball, since an external fire impinging on the vapor space of a pressure vessel is a common cause. However, it is not necessary for the liquid to be flammable to have a BLEVE occur. |
| **Bow Tie Model** | A risk diagram showing how various threats can lead to a loss of control of a hazard and allow this unsafe condition to develop into a number of undesired consequences. The diagram can also show all the barriers and degradation controls deployed. |
| **Conduct of Operations** | The embodiment of an organization's values and principles in management systems that are developed, implemented, and maintained to (1) structure operational tasks in a manner consistent with the organization's risk tolerance, (2) ensure that every task is performed deliberately and correctly, and (3) minimize variations in performance. |

**Distributed Control System** — A system which divides process control functions into specific areas interconnected by communications (normally data highways), to form a single entity. It is characterized by digital controllers and typically by central operation interfaces. Distributed control systems consist of subsystems that are functionally integrated but may be physically separated and remotely located from one another. Distributed control systems generally have at least one shared function within the system. This may be the controller, the communication link or the display device. All three of these functions maybe shared. A system of dividing plant or process control into several areas of responsibility, each managed by its own CPU, with the whole interconnected to form a single entity usually by communication buses of various kinds.

**Failure Modes and Effects Analysis** — A systematic method of evaluating an item or process to identify the ways in which it might potentially fail, and the effects of the mode of failure upon the performance of the item or process and on the surrounding environment and personnel.

**Hazard and Operability Study** — A systematic qualitative technique to identify process hazards and potential operating problems using a series of guide words to study process deviations. A HAZOP is used to question every part of a process to discover what deviations from the intention of the design can occur and what their causes and consequences may be. This is done systematically by applying suitable guide words. This is a systematic detailed review technique, for both batch and continuous plants, which can be applied to new or existing processes to identify hazards.

**Hazard Identification**

Part of the Hazard Identification and Risk Analysis (HIRA) method in which the material and energy hazards of the process, along with the siting and layout of the facility, are identified so that a risk analysis can be performed on potential incident scenarios.

**Hazard Identification and Risk Analysis**

Hazard Identification and Risk Analysis (HIRA): A collective term that encompasses all activities involved in identifying hazards and evaluating risk at facilities, throughout their life cycle, to make certain that risks to employees, the public, and/or the environment are consistently controlled within the organization's risk tolerance.

**Highly Managed Alarm**

An alarm belonging to a class with additional requirements (e.g., regulatory requirements) above general alarms.

**Human Machine Interface**

The means by which human interaction with the control system is accomplished

**Human Reliability Analysis**

A method used to evaluate whether system-required human actions, tasks, or jobs will be completed successfully within a required time period. Also used to determine the probability that no extraneous human actions detrimental to the system will be performed.

**Inspection, Testing and Preventive Maintenance**

Scheduled proactive maintenance activities intended to (1) assess the current condition and/or rate of degradation of equipment, (2) test the operation/functionality of equipment, and/or (3) prevent equipment failure by restoring equipment condition.

**Integrity Operating Window**
An Integrity Operating Window (IOW) is a set of limits used to determine the different variables that could affect the integrity and reliability of a process unit. An IOW is the set of limits under which a process, piece of equipment, or unit operation can operate safely. Working outside of IOWs may cause otherwise preventable damage or failure.

**Lagging Metric**
A retrospective set of metrics based on incidents that meet an established threshold of severity.

**Layer of Protection Analysis (LOPA)**
An approach that analyzes one incident scenario (cause-consequence pair) at a time, using predefined values for the initiating event frequency, independent protection layer failure probabilities, and consequence severity, in order to compare a scenario risk estimate to risk criteria for determining where additional risk reduction or more detailed analysis is needed. Scenarios are identified elsewhere, typically using a scenario-based hazard evaluation procedure such as a HAZOP Study.

**Leading Metric**
A forward-looking set of metrics that indicate the performance of the key work processes, operating discipline, or layers of protection that prevent incidents.

**Loss of Primary Containment**
An unplanned or uncontrolled release of material from primary containment, including non-toxic and non-flammable materials (e.g., steam, hot condensate, nitrogen, compressed $CO_2$ or compressed air).

**Management of Change**
A management system to identify, review, and approve all modifications to equipment, procedures, raw materials, and processing conditions, other than replacement in kind, prior to implementation to help ensure that changes to processes are properly analyzed (for example, for potential adverse impacts), documented, and communicated to employees affected.

| | |
|---|---|
| **Management of Organizational Change** | Management of organizational change (MOOC) is a framework for managing the effect of new business processes, changes in organizational structure or cultural changes within an enterprise. MOOC addresses the people side of change management. |
| **Normalization of Deviance** | A gradual erosion of standards of performance as a result of increased tolerance of nonconformance. |
| **Pressure Safety Valve** | A pressure relief device which is designed to reclose and prevent the further flow of fluid after normal conditions have been restored. |
| **Pre-Startup Safety Review** | A systematic and thorough check of a process prior to the introduction of a highly hazardous chemical to a process. The PSSR must confirm the following: Construction and equipment are in accordance with design specifications; Safety, operating, maintenance, and emergency procedures are in place and are adequate; A process hazard analysis has been performed for new facilities and recommendations have been resolved or implemented before startup, and modified facilities meet the management of change requirements; and training of each employee involved in operating a process has been completed. |
| **Process Hazard Analysis** | An organized effort to identify and evaluate hazards associated with processes and operations to enable their control. This review normally involves the use of qualitative techniques to identify and assess the significance of hazards. Conclusions and appropriate recommendations are developed. Occasionally, quantitative methods are used to help prioritize risk reduction. |
| **Process Safety Incident Database** | A database that is used to collect and record information from past process safety incidents. |

**Process Safety Management**
A management system that is focused on prevention of, preparedness for, mitigation of, response to, and restoration from catastrophic releases of chemicals or energy from a process associated with a facility.

**RAGAGEP**
"Recognized and generally accepted good engineering practice", a term originally used by OSHA, stems from the selection and application of appropriate engineering, operating, and maintenance knowledge when designing, operating and maintaining chemical facilities with the purpose of ensuring safety and preventing process safety incidents. It involves the application of engineering, operating or maintenance activities derived from engineering knowledge and industry experience based upon the evaluation and analyses of appropriate internal and external standards, applicable codes, technical reports, guidance, or recommended practices or documents of a similar nature. RAGAGEP can be derived from singular or multiple sources and will vary based upon individual facility processes, materials, service, and other engineering considerations.

**Risk Based Process Safety**
The Center for Chemical Process Safety's process safety management system approach that uses risk-based strategies and implementation tactics that are commensurate with the risk-based need for process safety activities, availability of resources, and existing process safety culture to design, correct, and improve process safety management activities.

**Reliability Centered Maintenance**

A systematic analysis approach for evaluating equipment failure impacts on system performance and determining specific strategies for managing the identified equipment failures. The failure management strategies may include preventive maintenance, predictive maintenance, inspections, testing, and/or one-time changes (e.g., design improvements, operational changes).

**Risk Based Inspection**

A risk assessment and management process that is focused on loss of containment of pressurized equipment in processing facilities, due to material deterioration. These risks are managed primarily through equipment inspection.

**Safe Operating Limits**

Limits established for critical process parameters, such as temperature, pressure, level, flow, or concentration, based on a combination of equipment design limits and the dynamics of the process.

**Safety Instrumented System**

A separate and independent combination of sensors, logic solvers, final elements, and support systems that are designed and managed to achieve a specified safety integrity level. A SIS may implement one or more Safety Instrumented Functions (SIFs).

**Safety Management System**

Comprehensive sets of policies, procedures, and practices designed to ensure that barriers to episodic incidents are in place, in use, and effective.

**Situational Awareness**

The conscious dynamic reflection on the situation by an individual. It provides dynamic orientation to the situation, the opportunity to reflect not only the past, present and future, but the potential features of the situation. The dynamic reflection contains logical-conceptual, imaginative, conscious, and unconscious components which enables individuals to develop mental models of external events.

| | |
|---|---|
| **Standard Operating Procedure** | Written, step by step instructions and information necessary to operate equipment, compiled in one document including operating instructions, process descriptions, operating limits, chemical hazards, and safety equipment requirements. |
| **Stop Work Authority** | A program designed to provide employees and contract workers with the responsibility and obligation to stop work when a perceived unsafe condition or behavior may result in an unwanted event. |
| **Subject Matter Expert** | A person who possesses a deep understanding of a particular subject. The subject in question can be anything, such as a job, function, process, piece of equipment, software solution, material, or historical information. Subject matter experts may have collected their knowledge through intensive levels of schooling, and/or through years of professional experience with the subject. |
| **Transient Operation HAZOP (TOH)** | A specialized HAZOP that focuses on hazards during transient operations such as commissioning startup, and shutdown. The TOH process centers on identification of required unit-specific activities (tasks) with a potential for an acute loss of containment and an in-depth review of the procedural controls necessary for safe and successful completion of those tasks. |
| **What-If Analysis** | A scenario-based hazard evaluation procedure using a brainstorming approach in which typically a team that includes one or more persons familiar with the subject process asks questions or voices concerns about what could go wrong, what consequences could ensue, and whether the existing safeguards are adequate. |

# ACKNOWLEDGMENTS

The American Institute of Chemical Engineers (AIChE) and the Center for Chemical Process Safety (CCPS) express their appreciation and gratitude to all members of the Guidelines for Managing Abnormal Situations Project Committee and their member companies for their generous efforts and technical contributions. The AIChE and CCPS also express their gratitude to the team of authors from Baker Engineering and Risk Consultants, Inc. (BakerRisk®) and the Abnormal Situation Management® Consortium. The AIChE and CCPS also thank the Abnormal Situation Management Consortium for providing partial funding for the publication of these *Guidelines*.

**Committee Members:**

| | |
|---|---|
| A. M. (Tony) Downes | Project Chair - Honeywell International, Inc. |
| Brian Farrell | CCPS Staff Consultant |
| Todd Aukerman | Lanxess |
| Felix Azenwi Fru | National Grid |
| Susan Behr | Braskem |
| Michelle Brown | FMC |
| Eddie Dalton | BASF |
| Seshu Dharmavaram | Air Products |
| SP Garg | GAIL (India) Ltd |
| Paul Gathright | Ascend Perf Materials |
| Amy Gay | Oxy |
| Nirupama Gopalaswami | Honeywell International, Inc. |
| Ghaffar Keshavarz | NOVA Chemicals |
| Alan Kobar | Dow |
| Rebecca Lipp | Marathon Petroleum |

| | |
|---|---|
| April Lovingood | Eastman |
| Marcus Miller | TapRoot |
| Assem Saleh | Petro Rabigh |
| Nicholas Sands | DuPont |
| Bryant Sartor | AdvanSix |
| Bruce Spencer | Marathon Petroleum |
| Todd Stauffer | exida |
| Andrew Trenchard | Honeywell (ASM Consortium representative) |
| Matthew Walters | Exponent |
| Stephanie Wardle | Cenovus Energy |

CCPS wishes to acknowledge the many contributions of the BakerRisk staff who contributed to this book, especially the writers listed here.

Larry O. Bowler

Michael P. Broadribb

Michael D. Moosemiller

Duane L. Rehmeyer

Roger C. Stokes

Editorial contributions in editing, layout, and assembly of the book were provided by Phyllis Whiteaker, BakerRisk document editor.

Before publication, all CCPS books are subjected to a thorough peer review process. CCPS gratefully acknowledges the thoughtful comments and suggestions of the peer reviewers. Their work enhanced the accuracy and clarity of these guidelines. Although the peer reviewers have provided many constructive comments and suggestions, they were not asked to endorse this book and were not shown the final manuscript before its release.

**Peer Reviewers:**

| | |
|---|---|
| Mohd Ibrahim Mohd Ashraf | Petronas |
| Palaniappan Chidambaram | DuPont Sustainable Solutions |
| Robert Johnson | Unwin Company |
| Caitlin Mullan | Ashland |
| Connor Murray | CNRL |
| Beverley Perozzo | NOVA Chemicals |

# PREFACE

Many adverse events have occurred in industry and elsewhere, due to abnormal situations that took place, or developed, but were not recognized in time or managed in a way that could have prevented the incident. Various tools and techniques, including complex automated control systems, are available to help manage such situations. However, such systems are not always effective if operators are not trained on them properly. By carefully considering how these abnormal situations might occur and by developing methods to identify, respond and manage them, the consequences that arise from them can be prevented or at least mitigated. This book examines such methodologies and management systems and provides a resource for operations and maintenance staff to be able to effectively handle abnormal events, as well as reduce the frequency and magnitude of process safety events.

The American Institute of Chemical Engineers (AIChE) has been closely involved with process safety and loss control issues in the chemical, petrochemical, and allied industries for more than four decades. AIChE publications and symposia have become information resources for those devoted to process safety and environmental protection.

AIChE created the Center for Chemical Process Safety (CCPS) in 1985 after significant chemical disasters in Mexico City, Mexico, and Bhopal, India. The CCPS is chartered to develop and disseminate technical information for use in the prevention of major chemical accidents. The center is supported by more than 200 chemical process industry sponsors that provide the necessary funding and professional guidance to its technical committees.

The major product of CCPS activities has been a series of guidelines to assist those implementing various elements of a process safety and risk management system. This book is part of that series.

CCPS strongly encourages companies around the globe to adopt and implement the recommendations contained within this book.

# DEDICATION

**John F. Murphy, PE, CCPSC**

CCPS and the members of the Guidelines for Managing Abnormal Situations subcommittee dedicate this book to Mr. John F. Murphy, PE, CCPSC. John is a CCPS Fellow, an AIChE Fellow, and has served CCPS and the AIChE Safety and Health Division for many years. John also served on the advisory board of Process Safety Progress, and has been a co-editor for over ten years at the time of this book's publication. Perhaps his most significant contribution was as a co-developer of the CCPS Fundamentals of Process Safety course that has provided thousands of industry professionals a firm grounding in process safety. John worked at Dow Chemical in multiple roles, including as leader of Dow's chemical reactivity hazards group.

His career also included five years as a Lead Investigator for the US Chemical Safety and Hazards Investigation Board (CSB). Among his contributions to the CSB was an exhaustive investigation of chemical reactivity hazard incidents spanning twenty years. These roles, among other process safety risk assessment roles, have helped industry improve the control of reactive hazards. Based on the critical importance of managing chemical reactivity hazards for preventing abnormal situations from escalating into major events, CCPS wishes to dedicate this book to recognize John's lifelong commitment to the cause of Process Safety.

Bruce Vaughen & Anil Gokhale

# 1 INTRODUCTION

## 1.1 PURPOSE AND SCOPE OF THE BOOK

This book provides resources for supervisors and operators/technicians in industrial processes, whose correct and timely intervention is often crucial, either to prevent abnormal situations from escalating into a major event, or to mitigate the consequences if an event occurs. Operations management and support services personnel (such as those in maintenance, engineering, and process safety), who read this book will be able to develop relevant training and support material to prevent or mitigate abnormal situations from occurring at their facility. This book includes the historical development of principles and procedures for managing abnormal situations as well as a summary and review of available resources for addressing them. It also provides guidance for management and engineers to develop appropriate training and procedures, and demonstrates how to institutionalize these into process operations.

Since many of the principles and practices of managing abnormal situations are transferable across industries, this book provides some guidance and suggestions on sharing knowledge and learning from a variety of industries and disciplines that are leaders in such management. With that in mind, example incidents (brief examples) and case studies (detailed studies) are included for front line staff as training aides.

As part of the development of this book, five online training modules were developed relating to abnormal situations. These training modules can be used by supervisors, plant engineers, and trainers to help train operating teams in diagnosing an abnormal situation. The modules allow the trainer to step through specific abnormal situations and discuss diagnosis, actions to be taken, learning, and relevance to their operation with the team members who are being trained. Details on how to access this material is provided in Appendix A.

## 1.2   WHAT ARE ABNORMAL SITUATIONS?

Abnormal situations in the process industries occur when conditions deviate from their normal state or operating range, such that basic process control systems are unable to restore normal conditions. Some examples of the early stages of an abnormal situation that could develop into a major incident, if not corrected, include:

- A tank level rises above safe operating limit because a control valve has seized.

- A tank level falls below safe operating limit because a level sensor has failed.

- A batch reactor temperature increases above safe limits because the wrong amount of one ingredient was added.

- Pressure starts to fall because a relief valve has lifted and has stuck open.

- The startup conditions are wrong because a key step was missed or delayed.

- A valve starts to close due to a leak of instrument air.

A typical intervention would be a response from a process operator, ranging from changing a set point on a control loop to bring the conditions back to normal, to a decision to carry out a controlled shutdown, or even to initiate an emergency shutdown of the process. Failure to respond adequately to an abnormal situation could lead to a worsening situation that could develop into a major incident, such as those described in Section 1.3.

Other commercial enterprises, such as the aviation and rail sectors of the transportation industry, also encounter abnormal situations, such as equipment failures, faulty warning signals or other unexpected safety events that must be managed. This book includes some example incidents and case studies from the aviation sector and examines their relevance to the chemical process industries. A useful example from the rail sector involves a runaway train at Lac-Mégantic (Transportation Safety Board of Canada Report 2014).

An Abnormal Situation, as defined by CCPS (CCPS Glossary):

"A disturbance in an industrial process with which the basic process control system of the process cannot cope."

A detailed definition of Abnormal Situation is included on the ASMC website (Abnormal Situation Management® Consortium) as follows:

- A disturbance or series of disturbances in a process that cause plant operations to deviate from their normal operating state.

- The nature of the abnormal situation may be of minimal or catastrophic consequence. It is the job of the operations team to identify the cause of the situation and execute compensatory or corrective actions in a timely and efficient manner.

- A disturbance may cause a reduction in production; in more serious cases it may endanger human life.

- Abnormal situations extend, develop, and change over time in the dynamic process control environments increasing the complexity of the intervention requirements.

User Centered Design Services (UCDS) Inc. describes abnormal situations as encompassing:

- "a range of events outside the 'normal' plant operating modes, i.e., trips, fires, explosions, toxic releases, human error, or just failing to reach planned targets" (UCDS website).

UCDS defines abnormal situation management as:

- A comprehensive process for improving performance, which addresses the entire plant population. It promotes effective utilization of all available resources—i.e., hardware, software, and people, to achieve safe and efficient operations.

Abnormal situations can be characterized as unplanned transitions or events that may occur during normal or routine operation, although they often take place during transient operations, such as startup, shutdown, switching to different products, or catalyst regeneration. These situations do not provide an opportunity for planning, training, and procedures to be developed to respond to them as they are occurring. Procedures and practices to manage abnormal situations should be established before an abnormal situation occurs. Operating personnel must be able to recognize the occurrence of an abnormal situation, identify its source, and take corrective action quickly and efficiently. However, in a dynamic process control environment, the situation can develop and change rapidly, making it difficult to determine the nature of the intervention that is required.

For optimal management of abnormal situations, the methodologies, tools, processes, and training should be established prior to the occurrence of such an event. This would help personnel recognize abnormal situations so they can either safely restore normal operations as quickly and efficiently as possible or take the process to a safe state, and subsequently to capture lessons learned to prevent similar situations from occurring in the future.

## 1.3     THE BUSINESS CASE FOR MANAGING ABNORMAL SITUATIONS

The consequences of a failure to provide the right intervention during an abnormal situation could ultimately lead to a major disaster such as a fire, explosion, toxic release, leading to environmental damage, injury and fatalities. Opportunities for early identification, allow corrective action to be taken before the abnormal situation escalates.

Examples of abnormal situations in order of increasing significance and consequences are included in this chapter. The relationship between abnormal situations and process safety events is provided in Figure 2.1 in Chapter 2.

- Loss of product quality.

- Reduction in operating efficiency/ increased processing costs.

- Deterioration of equipment (e.g., pump bearings, rates of corrosion), potentially leading to damage to the pressure envelope.

- Loss of containment of hazardous substances due to overpressure or overfilling, potentially leading to major events such as fire, explosion, toxic release, environmental damage, and fatalities.

Most of these consequences include a cost component such as lost production, equipment damage, injuries, or personnel costs including investigation and rectification work; as well as indirect consequences, such as damage to the environment, fines, and community complaints, even if the situations are managed well.

In addition, reputational damage tends to occur following a major incident, especially since most events are now broadcast widely via social media. Consequences of such incidents can negatively impact the surrounding community, to the point that some companies have shut down, as a result of major events. One example is the Philadelphia Energy Solutions refinery that was permanently closed after corrosion in a piping dead-leg led to a loss of containment, fire, and explosion in June 2019.

The benefits of adopting process safety principles are clearly outlined in the CCPS document, *The Business Case for Process Safety*, 4th edition (CCPS 2018b).

The cost to set up a management system that mitigates an Abnormal Situation is a fraction of the cost of the aftermath of process incidents. Trevor Kletz, renowned process safety expert and author, often remarked:

"If you think safety is expensive, try an accident" (Kletz/O'Connor PSC).

## 1.4 CONTENT AND ORGANIZATION OF THE BOOK

The chapter summaries provide an overview of the content and organization of this book, enabling the reader to locate a particular area of interest. Example incidents are embedded throughout the book to illustrate the points that are being made; some example incidents are used more than once where different aspects are relevant. Chapter 7 presents detailed Case Studies.

### Chapter 2 – Process Safety and Management of Abnormal Situations

Chapter 2 discusses why the management of abnormal situations is important to process safety and uses example incidents to illustrate the impact that abnormal situations, if unchecked, can have on operating facilities. It discusses some of the key elements of Risk Based Process Safety (RBPS) (CCPS 2007a) that are directly related to the management of abnormal situations, including: Hazard Identification and Risk Analysis (HIRA), Operating Procedures, Training and Performance Assurance, Asset Integrity and Reliability, Conduct of Operations, and Process Safety Culture.

The chapter makes the case for proactive management of abnormal situations; to avoid or mitigate an incident before it escalates into a more dangerous and costly event that can include downtime, lost production, equipment damage, injuries, and external / environmental and reputational damage. It reviews the consequences of a failure to manage abnormal situations, including an overview of some cases that demonstrate some of the principles discussed.

The chapter includes a more detailed discussion about the management of training for abnormal situations and the key issue of process safety culture. The latter includes the interaction between process safety culture, human factors, and the performance of individuals in stressful situations, including pressure to continue operations versus shutting down a facility. Process safety leadership is a strong influencer of culture, and this key issue is discussed in some depth.

### Chapter 3 – Abnormal Situations and Key Relevance to Process Plant Operations

Chapter 3 discusses subjects regarding management of abnormal situations that are relevant to process plant operations, explaining their relevance to operating personnel. It also includes plant design aspects, with separate focus points on new plant design and retrofits to existing plants. It includes a section on writing plant procedures and plant policies so that they incorporate the principles of managing abnormal situations.

Content topics explain basic theoretical knowledge, practical application of the theory in general, and its specific link and application to some of the 20 elements of Risk Based Process Safety that would apply to any industry. Practical strategies for implementation are also included, along with practices that impact the effectiveness of addressing abnormal situations to help improve safety, reliability, and efficiency of process operations.

This chapter also covers how to include and structure into existing procedures the best practices for managing abnormal situations, to help operating personnel make the right decisions during periods of abnormal situations. It also addresses how abnormal situations can occur during transient modes including startup, shutdown, or changes in operating modes and product grades, which can result in incidents.

### Chapter 4 – Education for Managing Abnormal Situations

Chapter 4 builds on Chapter 3 content to provide training on how to manage abnormal situations, with guidance on presenting the information to specific target populations. This includes advice, tools, and techniques for such training, both in the classroom and using DCS simulators or "digital twins" and E-Learning, and some computer-based training materials developed in conjunction with this book.

The primary target populations for this information are front line operators, operations managers, plant engineers, process safety engineers, technical experts, and others who could benefit from it. This chapter provides guidance to organizations in the process industries about organizing and structuring training on management of abnormal situations, including advice on trainer competencies, training programs, and assessment of required skills.

## Chapter 5 – Tools and Methods for Managing Abnormal Situations

Chapter 5 provides a brief summary of the relevant management tools that are currently available for recognition and management of abnormal situations; specifically, tools that illustrate the principles described in Chapter 3, and some worked examples of how to use them.

Hazard Identification and Risk Analysis is also addressed, including a list of scenarios that Process Safety Engineers can use during HIRAs to address abnormal situations.

## Chapter 6 – Continuous Improvement for Managing Abnormal Situations

Chapter 6 includes detailed descriptions of the available metrics related to managing abnormal situations that can be used for improvement, as well as the impact on incident investigations related to complex system issues, including lower-level incidents such as near-misses. The chapter also provides guidance on identifying the system and conditions that caused the incident and how to fix the system in order to prevent future incidents.

## Chapter 7 – Case Studies/Lessons Learned

Throughout the book, brief example incidents are used to illustrate specific issues in managing abnormal situations. Chapter 7 focuses on three case studies in more detail to illustrate topics discussed in earlier chapters.

The case studies address general as well as specific information regarding abnormal situations that have occurred both within and outside the process industry and how these events were handled from the perspective of management of abnormal situations. Both positive and negative examples are included in concise summaries.

# 2 PROCESS SAFETY AND MANAGEMENT OF ABNORMAL SITUATIONS

## 2.1 IMPACT ON PROCESS SAFETY

Process safety requires a disciplined framework for managing the integrity of operating systems and processes handling hazardous substances by applying good design principles, engineering, and operating practices. Therefore, process safety management systems focus on the prevention of, preparedness for, mitigation of, response to, and restoration from catastrophic releases of chemicals or energy from process facilities. At some companies, these management systems have been in place for many years and are generally credited with reducing major accident risk in the process industries. The risks can range from incidents involving toxic and flammable material releases resulting in toxic effects, fires, or explosions with potential impacts of harm to people (injuries, fatalities), to incidents that can cause harm to the environment, property damage, or production losses, and provide a conduit for adverse business publicity.

As discussed in Chapter 1, abnormal situations can be characterized as unplanned transitions or events that occur during normal and transient operations. For example, one or more process upsets that the automated control system (typically a distributed control system (DCS)) cannot correct can be the basis of an abnormal situation. This event may require intervention by operator(s) to augment the control system, to avoid a production/quality impact, or potentially, a serious process safety incident. Figure 2.1 illustrates the relationship between abnormal situations and the process safety incident categories in the CCPS and American National Standard Institute (ANSI)/American Petroleum (API) process safety metrics documents (CCPS 2018e; ANSI/API 2016).

Figure 2.1    Relationship of Abnormal Situations to Process Safety

Management of abnormal situations is inherently related to process safety performance and the avoidance of process safety incidents. The relationship in Figure 2.1 demonstrates that abnormal situations may progress from challenges to protection layers (barriers) to a minor incident, and ultimately to a major incident if the abnormal situation is not managed in a timely manner. Furthermore, since abnormal situations can occur without warning, the timescale of such situations does not permit activities such as planning, training, and procedure development, to respond to them as they occur. Consequently, process safety elements and principles should be applied to address abnormal situations *before* an incident occurs.

A number of process safety elements are key to the management of abnormal situations. The most important of these elements for safe operation are:

- **Hazard Identification and Risk Analysis (HIRA)** to identify, understand, and predict what abnormal situations can occur (CCPS 2007a, 2008b, 1999, 2010),

- **Operating Procedures** for transient and normal operation to ensure operations personnel have access to safe operating limits, consequences of deviation from safe limits, troubleshooting, and actions required to correct a deviation to prevent abnormal situations progressing to a process safety incident (CCPS 2007a, 1996),

- **Training and Performance Assurance** to build competency (knowledge, skill, ability) in operating procedures, control systems, recognizing warning signs, troubleshooting, and understanding process hazards to aid managing abnormal situations (CCPS 2007a, 2015, 2011a),

- **Asset Integrity and Reliability** to ensure that process equipment and control systems remain fit for purpose and reliable throughout their life to minimize challenges to protection layers (CCPS 2007a, 2017a, 2017c, 2007b),

- **Conduct of Operations** and **Operational Discipline** to ensure that all tasks including those essential for safe operation are performed reliably to minimize errors leading to abnormal situations (CCPS 2007a, 2011b, 2018f),

- **Process Safety Culture** to maintain the values and behaviors of a sound culture to deliver safe operations and improve human factors to help provide the conditions that support maximum performance of workers during abnormal situations (CCPS 2007a, 2018c, 2006, 2004), and

- **Incident Investigation** to learn from experience of prior abnormal situations and take action to strengthen management systems and process control to avoid and/or mitigate future abnormal situations (CCPS 2019).

Application of these and other process safety elements for managing abnormal situations is discussed in detail in Chapter 3.

## 2.2   THE CASE FOR POSITIVE MANAGEMENT OF ABNORMAL SITUATIONS

An abnormal situation typically starts with one or more operating parameters drifting outside normal limits that may impact product yield and quality. However, if this condition is not managed positively and quickly, the situation can rapidly escalate to a more dangerous and costly event that may include downtime, lost production, equipment damage, or injuries, as well as external property, environmental, and reputational damage.

Figure 2.2 (BakerRisk 2021) illustrates the concept of operating limits, showing the deviation of an operating parameter from "normal", through a "troubleshooting" zone, into an "emergency" zone. Once the emergency limit has been reached, an automated system might be available to shut down the process.

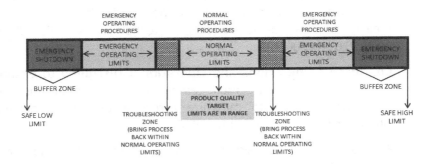

**Figure 2.2  Operating Ranges and Limits**

The speed with which the escalation outside normal limits occurs means that often only the control panel operator is available to respond if he/she even recognizes the problem. Technical and supervisory staff may have a greater understanding of the process technology but may be unreachable during shift hours or have obligations in other areas of the plant. Therefore, the ability of the control panel operator to troubleshoot and correctly diagnose the problem is paramount to prevent, or at least mitigate, a process safety incident. This is especially important if the abnormal situation can escalate to a complete loss of control and a serious process safety emergency. Additionally, the level of sophistication of modern process control systems including DCS and advanced process control (APC) systems significantly limits opportunities for control panel operators to regularly experience and practice process interventions.

The application of process safety elements, principles, tools, and techniques provides an opportunity to manage an abnormal situation before it escalates. It is vitally important that operations personnel be taught how to intervene during abnormal situations and react accordingly. In particular, consideration should be given to the use of simulators for training operators in abnormal situations during transient (e.g., startup, shutdown) and normal operations (Chemical Engineering 2016). All abnormal situations should be recorded and shared with operations personnel, including lessons learned and actions taken to improve managing of similar situations in the future.

## 2.3    ADVERSE OUTCOMES OF ABNORMAL SITUATIONS

The frequent occurrence of abnormal situations increases the likelihood of process safety incidents at a facility. An abnormal situation often occurs as an early step in a series of events that lead to serious incidents.

Industry and insurance company surveys have indicated that the cost of the consequences of process upsets and other unplanned events can range from $100,000 to many millions of dollars. While the cost of equipment damage may be claimable, depending on insurance coverage, the actual cost to companies is likely significantly higher due to policy deductibles, business interruption that may not be claimable, and possible reputational damage.

A 2020 insurance study analyzed 137 incidents between 1996 and 2019 that resulted in major losses (> $50 million) in the onshore oil, gas, and petrochemical industries (Jarvis & Goddard 2020). Figure 2.3 from the study shows the breakdown of cause of loss between mechanical integrity failure, unsafe maintenance, and operations.

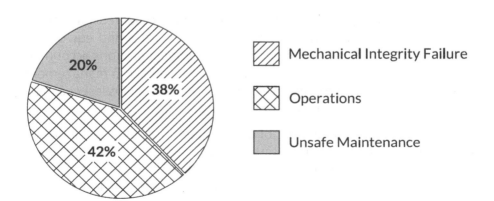

**Figure 2.3  Breakdown by Loss Type**

Operations losses in the study accounted for 42% of the total (57 losses). The loss type classification differs somewhat from the definition of an abnormal situation. The definition used in the insurance study was:

"Everything else not classified as mechanical integrity failure or unsafe maintenance, but mostly concerning non-routine, infrequent or abnormal or unplanned operations sometimes collectively referred to as transient operations."

Figure 2.4 from the study shows the classification of the operations losses by operating mode.

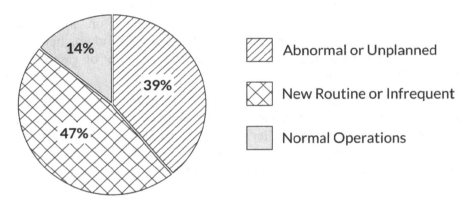

**Figure 2.4  Breakdown of Operations Losses by Operating Mode**

Non-routine/infrequent planned operations accounted for 47% of the total (27 losses), and were dominated by startup and equipment switching operations, while abnormal/unplanned operations accounted for 39% (22 losses), including 11 utility failures, and alarm management was often a contributory factor. The remaining 14% (8 losses) occurred during normal operations (Jarvis & Goddard 2020).

A disproportionate number of process safety incidents occur during transient operations, such as startup, shutdown, catalyst regeneration, decoking, and other non-routine operations (CSB 2018). It has been determined that approximately 50% of process safety incidents occur during these transient operations, despite these representing less than 10% of the operating life of a typical process facility (CCPS 1995; Duguid 1998a, 1998b, 1998c). Transient operations involve non-routine procedures and can result in unexpected and unusual situations.

Furthermore, improvements in reliability and extended intervals between process unit turnarounds inevitably result in operations personnel having less familiarity with non-routine operations.

Example Incident 2.1 and Example Incident 2.2 (Broadribb 2003, CSB 2011) involving transient operations that led to major incidents demonstrate the importance of understanding the causes of abnormal situations and establishing effective management strategies.

**Example Incident 2.1 – BP Amoco Polymers, Augusta, GA, 2001**

The Polyamide Unit produces a high-performance nylon by polycondensation in two preheaters at high pressure. Water flashes off in the reactor at high temperature; the liquid-vapor product is then fed to the extruder where polymer is stranded (see Figure 2.5).

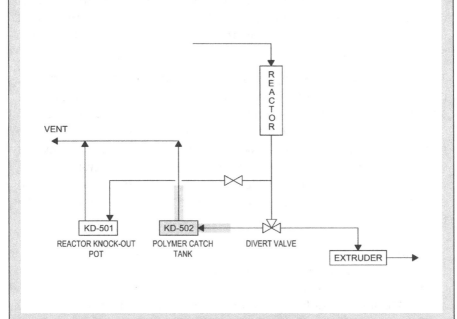

**Figure 2.5  Polyamide Unit Process Flow Diagram**

**Example incident 2.1 – BP Amoco Polymers (*cont.*)**

BP Amoco had shut down the Polyamide Unit for maintenance, and restarted it afterward, which required diversion of off-specification pre-polymer to the polymer catch tank. Just before the feed was to be switched to the extruder, operators found that the extruder screws would not turn, despite several attempts. This extended the normal period of pre-polymer diversion, and the unit was shut down for cleaning. However, during a solvent flush, the polymer catch tank [KD-502] developed a flange leak, and the flush was diverted to the reactor knockout (KO) pot [KD-501], followed by a water wash. During the water wash, operators did not realize that the relief valve on the reactor KO pot had lifted to relieve steam pressure to the blowdown tank, even though a blowdown tank temperature indicator alarmed. Abnormal pressure in the reactor could not be relieved to the polymer catch tank, so it was relieved to the reactor KO pot. The common vent line from the polymer catch tank and reactor KO pot was left open to the vent scrubber.

Two operators made energy isolation preparations for opening the polymer catch tank to remove waste polymer deposited during the aborted startup. The maintenance technician, assisted by the two operators, had just removed about half of the 44 bolts retaining the polymer catch tank end plate, when the end plate suddenly blew off violently, fatally striking the three employees and forcibly ejecting vessel internals and hot molten polymer. Six minutes later, an explosion occurred when oil from damaged hot oil lines ignited, and a fire ensued.

The abnormal situations were:
  i.   Extended pre-polymer disposal,
  ii.  Additional polymer disposal during shutdown,
  iii. Absence of cooling from the solvent flush and water wash,
  iv.  Immediate attempt to remove waste from the polymer
These conditions were unique in the history of the unit.

**Example Incident 2.1 – BP Amoco Polymers (*cont.*)**

The molten pre-polymer entering the polymer catch tank was about 620 °F (327 °C), resulting in thermal decomposition in the absence of cooling. This decomposition caused polymer foaming and expansion, which flowed into the unheated connected piping and solidified. The vessel relief, vent, and drain lines were found to be full of solid polymer when the unit was dismantled, which explained the presence of significant pressure in the polymer catch tank that was not apparent from checking pressure gauges and opening drain valves. The blocked locations are shown in the shaded areas in Figure 2.5.

Lessons Learned in relation to abnormal situations:

1) For management/engineers:
   - Polymer decomposition was not fully understood, identified, or adequately discussed in HIRA studies and operating procedures.

   - The polymer catch tank design was unsuitable for the dirty service of waste polymer.

2) For supervisors / operators / technicians:
   - Operations and maintenance personnel were unaware of hazards regarding polymer thermal decomposition.

   - Absence of a formal maintenance procedure for cleaning the polymer catch tank.

   - A startup procedure to test the extruder was not followed.

   - Operational readiness review (aka pre-startup safety review) should be conducted before every startup.

**Example Incident 2.2 – Bayer Crop Science Plant, Institute, WV, 2008**

The Methomyl Unit produces a chemical used in the manufacture of a pesticide, Larvin®. After an extended shutdown to upgrade the control system and replace the residue treater pressure vessel, an incident occurred during startup. A runaway chemical reaction occurred inside the new residue treater, overwhelming the relief system, and causing the vessel to explode violently, resulting in two fatalities. Highly flammable solvent sprayed from the vessel and immediately ignited, causing an intense fire that burned for more than 4 hours. Six volunteer firefighters and two contractors working nearby were treated for possible toxic chemical exposure. More than 40,000 residents, including students at West Virginia State University, were advised to shelter in place for more than three hours as a precaution. Police closed the interstate highway and roads near the facility due to drifting smoke.

The abnormal situations caused the runaway reaction:

i.   Malfunctioning equipment, including the new DCS that was not tested and calibrated, prevented achievement of correct operating conditions in the crystallizers and solvent recovery equipment,

ii.  Methomyl-solvent mixture was fed to the residue treater before it was pre-filled with solvent and heated to the minimum safe operating temperature,

iii. Methomyl that had not crystallized due to equipment problems greatly increased the methomyl concentration in the residue treater.

A newly installed relay malfunctioned, tripping both centrifuges when operators attempted to run both at the same time as per normal procedure. While starting the unit, staff discovered a missing valve on a solvent drip line and a broken cooling water valve. These two equipment installation problems directly contributed to the overconcentration of methomyl.

**Example Incident 2.2 – Bayer Crop Science Plant, (*cont.*)**

Lessons Learned in relation to abnormal situations:

1) For management / engineers:
   - Overly complex operating procedures

   - Inadequate operator training on the newly installed DCS

   - Temporary changes not evaluated

   - Malfunctioning or missing equipment (faulty new relay caused both centrifuges to trip, solvent drip line was missing a valve)

   - Insufficient technical expertise available in the control room during the restart

   - Operational readiness review was inadequate

2) For supervisors / operators / technicians:
   - Deviation from written operating procedure (several required steps were not completed)

   - Safety-critical equipment bypassed (interlocks on residue treater) or not operable (toxic gas monitoring system not in service)

   - Misaligned valves

Example Incident 2.1 and Example Incident 2.2 – (Broadribb, CSB) involved transient operations of startup and shutdown. However, abnormal situations can also occur during normal operation, as Example Incident 2.3 (HSE 1997) illustrates. More details of Example Incident 2.3 are also provided in Chapter 7.2.

**Example Incident 2.3 – Texaco Refinery, Milford Haven, Wales, 1994**

Significant process upsets occurred when a lightning strike caused disturbances to the distillation, alkylation, fluidized catalytic cracking unit (FCCU) and other process units. All units except the FCCU were shut down because of the lighting strike.

An explosion occurred approximately 5 hours later and released about 20 tonnes (44,000 lbs.) of flammable hydrocarbons from the outlet pipe of the FCCU flare knockout (KO) drum. Due to explosion damage to the flare system, the subsequent fires were allowed to safely burn out two days later. The damage repair cost about $65 million, and the refinery was shut down for two months, with additional business interruption to the damaged area.

The cause of the explosion was due to several abnormal situations arising from a series of events and failures in management, equipment, and control systems during the process upsets:

- A control valve on the outlet of the debutanizer closed automatically shortly after the lightning struck, and although the control system indicated that the valve had re-opened, it remained closed. This allowed the debutanizer to fill up and pressurize, causing it to vent via a safety valve to the flare KO drum on several occasions.

- In order to relieve the pressure on the debutanizer, operators vented the overheads drum to a compressor surge drum, which caused further problems, as the wet gas compressor then tripped.

- Operators focused their attention on the compressor and made several attempts to restart it. The net effect was that a significant quantity of hydrocarbons was fed to the flare KO drum, such that it was overfilled. Liquid was carried forward to the flare line that was only designed for vapor. Weakened by corrosion, the flare line fractured, which released hydrocarbons that subsequently ignited, resulting in a vapor cloud explosion (VCE).

**Example Incident 2.3 – Texaco Refinery, Milford Haven (*cont.*)**

- Operators missed key information, such as the buildup of liquid in the KO drum, which could have prevented the explosion. However, a previous modification that involved the removal of the automatic pump-out system limited their options.

- For several hours after the lightning struck, operators were overwhelmed by a flood of 2040 alarms (at an estimated rate of one every two to three seconds), all of which were designated 'high' priority. Many were only informative, but the existing alarm design did not separately prioritize and display safety-critical alarms. Alarms were annunciated faster than the operators could recognize, acknowledge, and respond to them, including 275 alarms in the final 11 minutes prior to the explosion.

Lessons Learned in relation to abnormal situations:

1) For management / engineers:
   - Relief systems should be designed to handle worst-case abnormal situations.

   - Safety critical alarms should be distinguishable from other operational alarms.

   - The number of alarms should be limited to allow effective monitoring by operators. A "first-out" alarm system can be a useful design feature in these circumstances. See Chapter 5 on tools and methods.

   - Control panel graphics should include a process overview screen to help with troubleshooting.

   - Modifications (removal of pump-out system) should consider abnormal situations.

   - A plant safety system must be robust enough to handle situations where a human response to an alarm is not enough to mitigate the issue.

---

**Example Incident 2.3 – Texaco Refinery, Milford Haven (*cont.*)**

2) For supervisors / operators / technicians:
- Control systems, including each control valve, should be tested prior to all startups as part of an operational readiness review. As a minimum, safety-critical systems should be tested after plant trips.

- Operators must remain alert to recognize abnormal situations, such as high levels in vessels.

---

These example incidents illustrate that abnormal situations can result in major, tragic, and costly process safety incidents. However, most incidents are preventable if the abnormal situation is recognized, diagnosed correctly, and corrected in a timely manner. Therefore, it is essential that operations personnel have the knowledge, skills, and abilities to manage abnormal situations.

## 2.4   IMPORTANCE OF TRAINING FOR ABNORMAL SITUATIONS

An abnormal situation is often recognized as an abnormal occurrence, but it may not always be recognized as a potential process safety issue. In order to prevent these types of incidents from occurring, whether the abnormal situations occur during transient or normal operations, it is imperative that companies understand the hazards, provide workers with appropriate training, and have in place and enforce robust process safety policies and procedures for all hazardous operations, including startups and shutdowns. These policies and procedures should address all elements of process safety (CCPS 2007a) and human factors (CCPS 2006, 2004), and specifically provide guidance indicating clearly that abnormal situations may have a process safety component to them and are not just operating difficulties.

If properly implemented, the RBPS *Operating Procedures* element includes the safe operating limits, consequences of deviation from safe limits, and the actions required to correct a deviation (CCPS 2007a). However, with new technology and the increasing complexity of some integrated process plants, imagining all the possible causes of an upset is difficult. If not immediately identified and correctly resolved, a process upset can develop and change over time in a dynamic process control situation, further complicating intervention requirements. Therefore, the 'actions to correct deviations' may or may not address the actual cause of the abnormal situation and guide the operator(s) in executing compensatory or corrective actions in a timely and efficient manner.

Extensive training, especially with a DCS simulator or digital twin, can improve the performance of the control panel operator (Chemical Engineering 2016). The science dedicated to improving human performance in relation to managing abnormal situations, includes:

- Operator situational awareness.

- Real-time support to the control panel operator.

- Operator information and communication systems.

- Human Machine Interface (HMI) design and ergonomic control rooms.

- Alarm management and operator response to multiple alarms.

- Early detection of process upsets.

Further information on managing abnormal situations is provided in Chapter 5.

## 2.5    SAFETY CULTURE AND THE MANAGEMENT OF ABNORMAL SITUATIONS

Abnormal situations are process upsets that the control system cannot correct properly, thus requiring operator intervention. While some improvements to the control systems may reduce abnormal situations, improvements in the reliability of human performance requires focus on more than technology. The ability of operators to effectively prevent and respond to abnormal situations is a key element in many plants and reduces the risk of such situations developing into process safety incidents. Human factors, such as situational awareness, Human Machine Interface, operator overload, fatigue, stress, and decision bias, can present issues that impair human performance. Additionally, operators must also be able to recall their training and adapt quickly and responsibly when things go wrong.

In addition, during times of transient conditions, such as unit startup, the risk of operators having conflicting understandings of the state of the process unit is greater. Leadership should ensure that experienced, technically trained personnel supervise and support operators during unit startups and shutdowns, and that effective communication and feedback is essential to establish and maintain a mutual understanding of the process unit and its expected future state.

Some cultural considerations related to managing abnormal situations can have an influence on human factors. Corporate and facility leadership can drive or limit the organization's safety culture, through words or actions. For example, if leadership seeks to assign blame for process upsets and incidents, the workforce will be unlikely to report events that could otherwise be learning opportunities and serve to reduce or mitigate future abnormal situations. Leadership must also put safety before production and profit. For instance, there should be no pressure on operators to continue operations, if shutdown is the correct response to an abnormal situation ("Stop Work Authority").

Abnormal situations introduce stress, and operators under stress can make poor decisions, which then exacerbate the situation. How companies prepare and equip their operators to deal with these problematic and stressful situations is critical to ensuring the return of the unit to a safe state. Often, process safety incidents are a result of organizations failing in this area.

Similarly, Conduct of Operations (COO) and Operational Discipline (OD) are closely tied to an organization's Process Safety Culture (CCPS 2007a, 2011b). If leadership enforces high standards, then a robust COO will ensure that operational tasks, such as established management systems and procedures, are executed in a deliberate and structured manner. Furthermore, OD is associated with the organizational and individual behaviors, and will dictate how well the management systems and procedures are applied.

A strong and positive process safety culture should ensure that operational tasks, such as operating and maintenance procedures, safe work practices, and shift handover communication, are followed routinely and diligently. Consequently, if operations personnel perform their duties with alertness, due thought, full knowledge, sound judgment, pride, and accountability, the impact of abnormal situations should be minimized, and serious process safety incidents prevented or mitigated.

This requires key personnel with leadership skills, and strong management systems, practical organizations, and communications systems. For example:

- Operating procedures (containing safe operating limits, consequences of deviation from safe limits, troubleshooting, and actions required to correct a deviation must be up-to-date, accurate, easy to use and understand, and readily accessible.

- Organizations should avoid complexity by having a limited number of vertical management layers and departments, to avoid interface issues and poor communication that could make it more difficult to react to a dynamic abnormal situation.

- Operators and their supervisors should be empowered to shut down the plant in the event of a serious abnormal situation, if insufficient time is available to escalate the problem within a hierarchical organization.

- Management should drive a 'no blame' culture within the plant and commend actions that prevent or mitigate potential process safety incidents.

In summary, most abnormal situations that result in process safety incidents also involve safety culture influences, including COO and OD issues (CCPS 2007a, 2011b). Every country develops its own culture, and its people develop habits, norms, and values that differ from other countries. National culture is often a good predictor of attitudes, behaviors, and performance in the workplace. For example, in some cultures, operators are likely to escalate an abnormal situation problem to the senior manager or superintendent, rather than take action, especially if it involves plant shutdown or stop work. However, in other cultures, workers are more likely to follow written policies, practices and procedures that may or may not satisfactorily address abnormal situations. Within Europe, for example, there are various sub-cultures with different approaches to an abnormal situation (EU-OSHA 2013). Similarly, in the USA there are sub-culture differences between states. For example, in some facilities, operators may have a 'can do' attitude and may not often refer to written operating procedures, while in other locations, operators may be more likely to diligently follow detailed instructions. Certain cultures are hesitant to contradict those in charge, even where all the data points to something being wrong.

International companies should formulate different plans for different country cultures. Managers should understand the local culture, try to adapt their leadership behavior to that preferred in the host country, and stimulate an inclusive working environment in which people from diverse backgrounds feel respected and recognized. Workforce training should focus on competencies that increase intercultural effectiveness at all levels.

Organizations wanting to improve their management of abnormal situations should start with an honest, straightforward assessment of their existing culture. Successful cultural change requires effective communication and reinforcement of expectations of new attitudes and behaviors. For example, leadership may wish to empower employees to stop work if it may be unsafe to continue operation. Cultural change takes time and leadership may need to communicate repeatedly the message. However, if over time, these new attitudes and behaviors demonstrate successful results, the workforce will recognize and appreciate the resulting successes (CCPS 2007a, 2018c, 2006, 2004).

# 3 ABNORMAL SITUATIONS AND KEY RELEVANCE TO PROCESS PLANT OPERATIONS

This chapter discusses focus areas for the management of abnormal situations and explains their relevance to process operating personnel. It includes plant design aspects, new technologies, operating modes during which abnormal situations can occur. The chapter also provides the links between abnormal situation management and CCPS' Risk Based Process Safety elements. The chapter discusses procedures for managing abnormal situations. Examples of abnormal situations encountered in a variety of example incidents are also included, and the associated lessons learned can be especially valuable for sharing with frontline supervisors and operators.

## 3.1 FOCUS AREAS FOR ABNORMAL SITUATION MANAGEMENT

Several areas relevant to the management of abnormal situations have been identified by the Abnormal Situation Management® Consortium (ASMC) as research areas, as summarized.

### 3.1.1 ASM Research Areas

The ASMC refers to seven areas of research concerning abnormal situations, recognizing that the connection between the system and the human, as well as human strengths and limitations, must be understood. A detailed understanding of these focus areas, and others as outlined in Section 3.1.2, is required for management of abnormal situations at the plant level. Most of these areas have a direct link to key elements of Risk Based Process Safety (CCPS 2007a), as introduced in Section 2.1, the relevance of which is detailed in Section 3.3.

The seven ASM® Consortium research areas are:

1) <u>Understanding Abnormal Situations</u>: This area focuses on issues that can lead to a better understanding of common root causes of incidents, including measuring, reporting, analyzing, and communicating the causes and effects associated with abnormal situations. These issues vary widely but can provide insight to reduce future abnormal situations, and to prepare operations teams to more quickly recognize, efficiently respond, and accurately handle the abnormal situations that do occur.

2) <u>Organizational Roles, Responsibilities and Work Processes</u>: This area focuses on management practices that influence the organizational culture, work processes, staff roles and responsibilities, and valued behaviors as they relate to abnormal situations. The organizational structure can affect the ability of the operations team to prevent and/or respond appropriately to abnormal situations.

3) <u>Knowledge and Skill Development</u>: This area focuses on the impact of training and skill development, in anticipating and coping with abnormal situations. Knowledge and skill development and maintenance of a competent workforce requires a continuous learning environment to be established. Ideally, the training and skill development content is based upon a needs analysis that defines the minimum acceptable knowledge, skills, and abilities to operate the chemical processes safely. It should include structured training on management of abnormal situations. The training and skill development must include maintenance and instrument technicians so personnel can safely maintain the equipment and assist in troubleshooting abnormal situations associated with equipment faults.

4) <u>Communications</u>: This area focuses on communications issues among plant personnel and the use of information technology under normal, abnormal, and emergency situations. This includes communication within and between different shift teams and situational dialog among plant personnel and explores opportunities to use information technology that improves site-wide coordination in a crisis. Although many abnormal situations are local to a specific area or processing unit at the plant, others

may have a site-wide effect. For example, an upset in a central utility unit or waste treatment unit may require rapid communication to all plant site units, resulting in quick response actions.

5) <u>Procedures</u>: This area focuses on all aspects of operational and task procedures used to accomplish important activities at an industrial site, particularly during startup and shutdown. This includes all aspects of procedures used, such as accessibility, accuracy, clarity, and policy compliance. A thorough assessment is needed for any procedural or instrumented process control deficiencies that do not address the possibility of an abnormal situation occurring. Section 3.3 and Chapter 5 will address procedures in more detail.

6) <u>Work Environment</u>: This category focuses on the impact of the control room and control building, especially workplace and ergonomic design factors that could lead to impeded performance of personnel during both normal operation and abnormal situations.

   a. Some examples of environmental factors on abnormal situations include tight spaces, noise, lack of visibility, time of day, access to safety gear, breakroom locations, or egress routes.

   b. An ergonomic assessment of the physical layout and various workspace conditions should be beneficial in identifying factors that could limit response to abnormal situations.

7) <u>Process Monitoring Control and Support Applications</u>: This area focuses on automation technologies for effective operations such as effective design, deployment, and maintenance of hardware and software platforms that support process monitoring, and control and support for effective operations. It includes issues such as alarm flooding and the Human Machine Interface (HMI) in abnormal situations.

## 3.1.2　Additional Focus Areas

Other key areas that require consideration for the management of abnormal situations include:

### 3.1.2.1　Instrument Failures

Potential failure modes should be identified and appropriate diagnostic/ troubleshooting skills, tools, and techniques, along with associated training should be developed for operators to handle situations including:

- Valve Failures:　A faulty or failed automatic valve can contribute to abnormal situations. Additionally, the possibility of an automatic valve not failing to a safe position during an electrical or pneumatic supply loss should be discussed and understood by the plant personnel.

- Sensor Failures:　A faulty or failed sensor can contribute to abnormal situations. In addition, operators should be aware of potential situations when sensors and instruments are off-line for calibration or repair.

- I/O Card failures:　Input and Output card failures in the process control network can occur and are often difficult to quickly diagnose and address.

- Bypassing alarms and trips:　There are occasions when it may be necessary to override such systems temporarily, for example, when a sensor fails. If not properly managed, however, this can either be the direct cause of an abnormal situation or remove a layer of defense if an abnormal situation arises from another cause. Good PSM systems include a rigorous management procedure to control alarm suppression and interlock bypassing, which is further discussed in Chapter 5.

Further details are provided in Chapter 4, Education for Managing Abnormal Situations.

### 3.1.2.2　Services Failure Including Power Blackout

Service failure includes situations in which one or more of the services (including electricity, steam, air, water, inert gas) is lost due to an outage or other unforeseen reason. During a power blackout, although an uninterruptible power supply (UPS) usually allows the DCS to continue for a limited time, it does not always work as intended. Guidelines and/or training for continued operation and/or safe shutdown in this circumstance should be developed. Further aspects are discussed under 3.4.2.3.

The example described in Example Incident 3.1 involves a power failure just to the process control system and the unforeseen consequences of restoring the supply.

---

**Example Incident 3.1 – Control System Power Failure**

A batch process using toxic and flammable chlorocarbons suffered a failure of power supply to the DCS. The process continued to operate safely without the DCS for a short time, since the system was set up so the controls would fail to a safe position. The reactor agitator control systems went to a stay-put mode, so the reactants continued to be mixed and the exothermic reactions were in control. When power to the DCS was restored, the operators then turned it back on, which forced all control parameters and variables to their initialization positions. This caused a number of problems, including a zero-speed for the agitators. As a result, the plant experienced a near-miss, as a chemical stop agent had to be used to kill the reaction in order to prevent a runaway reaction.

Lessons learned in relation to abnormal situation management:

- Abnormal Situation Recognition: Operating teams must be aware of the failure modes of equipment due to loss of services (e.g., power, steam, instrument air). Failure modes due to loss of services (power, steam, instrument air, or other utilities) should be identified during the risk assessment and be understood by designers and operating/maintenance teams.

- Procedures: Should be available, and training should be provided to handle such failures. Training should include failure of one service that can lead to cascading failure of other services (e.g., power, steam, air, water).

### 3.1.2.3   Emerging Technologies and Complex Systems

New technologies, including smart systems, connected devices, machine learning, and artificial intelligence can add to the safe and reliable operation of a process. However, these can also cause issues for operating teams dealing with an abnormal situation, as the systems may have inherent, unknown modes and patterns of failure. Since such systems can provide many diagnostic messages and fault indications that might be difficult to interpret, they should be carefully reviewed at the design stage to determine which messages and faults should be sent to the operator. It is recommended that new technologies are trialed in low-risk systems applying the MOC process so that operators and maintenance staff become familiar with them before they are applied to more safety-critical operations. One example is that of wireless technology that was initially used only for process monitoring but is now being used for control. However, wireless systems are not currently recommended for use on safety-critical systems.

Many new technologies increase the risk of a cyber-attack, whereby multiple barriers could be impacted simultaneously. Cyber issues are outside the scope of this book and are the subject of a separate CCPS project: "Managing Cybersecurity - A Risk Based Approach Building on the Process Safety Framework". Additionally, a history of cyber incidents associated with industrial control systems can be found in "History of Cyber Incidents and Threats to Industrial Control Systems" (Hemsley & Fisher 2018).

For guidance on safe process control design and management, CCPS has published *Guidelines for Safe Automation of Chemical Processes*, $2^{nd}$ edition (CCPS 2017c).

## 3.2   ABNORMAL SITUATIONS AFFECTING PROCESS PLANT OPERATIONS

Process plant personnel, especially frontline personnel, are trained on normal plant operating conditions and associated procedures to adjust, modify, and even mitigate conditions that are outside of the normal operating range of the process. Recognizing normal process deviations is one of their primary responsibilities and one that trained operators do well. However, process plants do not always maintain normal conditions and situations may arise that are not normal or have not been previously experienced. Some of these situations may have been anticipated, and procedures or automated systems are in place to deal with them; others may not be anticipated, and plant personnel will have to troubleshoot and resolve the issues using their skills, knowledge, and experience. Many of these abnormal process situations occur during startups, unanticipated interruptions, or when conducting non-standard operating tasks.

The design of many process plants incorporates complex and sophisticated process control systems to keep the plant running in optimum condition and to protect the safety of the plant and personnel. In addition, the plants should also follow Recognized And Generally Accepted Good Engineering Practices (RAGAGEP). Nevertheless, the plants can be subject to periodic failures or upset situations that are sometimes not easily recognized or controlled. These abnormal situations may not be immediately obvious and are sometimes undetected or overlooked; but if a situation should develop into an unwanted event, the consequences could be significant. Therefore, recognizing and managing these situations is vital to maintaining a stable and safe process.

Situation Awareness, which is sometimes referred to as Situational Awareness (SA), is an area of human factors that is highly relevant to abnormal situations. The topic encompasses how humans interact with complex systems, including perception of a situation, understanding what is going on and predicting how the status will change in the near future. One definition for SA can be found in Stanton (Stanton et al 2001) and is as follows:

> *"Situational awareness is the conscious dynamic reflection on the situation by an individual. It provides dynamic orientation to the situation, the opportunity to reflect not only the past, present and future, but the potential features of the situation. The dynamic reflection contains logical-conceptual, imaginative, conscious and unconscious components which enables individuals to develop mental models of external events." (Bedny & Meister, 1999)*

History indicates that abnormal situations that are not immediately recognized or effectively addressed can escalate at a facility, resulting in one or more of the following:

- Chemical release or fire leading to injuries or fatalities
- Equipment and property damage
- Environmental impact or non-compliance
- Business interruption
- Loss of community confidence/ reputational damage

Two often-referenced abnormal situation incidents are Union Carbide's chemical release in Bhopal, India in 1984 and the BP Texas City refinery isomerization unit explosion in 2005 (CCPS 2008a). As summarized in Example Incident 3.2 and Example Incident 3.3, both events included significant precursors that were allowed to continue, resulting in abnormal situations that led to major consequences.

---

**Example Incident 3.2 – Union Carbide, Bhopal, India 1984**

On December 3, 1984, the Union Carbide plant located about 3 to 4 miles (5 to 7 km) outside the center of the city of Bhopal, India accidentally released into the atmosphere approximately 40 metric tons of methyl isocyanate (MIC), an intermediate chemical used in production of carbaryl (a pesticide). The incident resulted in the fatalities of approximately 4,000 people living near the plant and the subsequent fatality of over 16,000, as well as injuries in 200,000 and genetic mutations to affect several generations of offspring (Gupta 2004).

The incident originated from one of the three MIC storage tanks, which were refrigerated, partially buried, and equipped with relief valves that discharged to a flare tower through a caustic soda scrubber. At the time of the accident, the MIC plant had been shut down for over a month, but carbaryl production was allowed to continue, using the inventory left in the storage tanks.

**Example Incident 3.2 – Union Carbide, Bhopal, India 1984 – (*cont.*)**

The immediate cause of the release was an exothermic reaction due to water entering one of the storage tanks. The plant had been designed with a refrigeration system and pressure relief system, to address the consequences of precisely that type of event. However, at the time of the accident the refrigeration system had been shut down for approximately 6 months to save money, the relief system, caustic soda scrubber was shut down for maintenance, and the flare had been shut down, awaiting replacement of corroded piping.

During the subsequent investigation, numerous other mechanical and operational issues were also observed: i.e., unreliable pressure gauges, failure of the high temperature alarm to activate, a storage tank filled above its recommended capacity, an emergency reserve MIC tank that already contained some inventory, and a safety water curtain that was designed to absorb escaping MIC vapors but had insufficient reach.

The release was first detected by plant operators when they experienced a burning sensation in their eyes and informed their supervisor, who failed to take immediate action. For over two hours, the chemical release formed a ground-level cloud that spread downwind into the neighboring community, with catastrophic results.

---

**Example Incident 3.2 – Union Carbide, Bhopal, India 1984 – (*cont.*)**

Lessons learned in relation to abnormal situation management

- Abnormal situation recognition: No Management of Change (MOC) was in place to approve the plant operating conditions during the accident. Abnormal situation status was either not recognized or more likely was accepted without a risk analysis.

- Organizational roles and work processes: Personnel response to the accident was inadequate and lacked direction from the supervisor. Plant site culture appeared to accept operating without critical safeguards functioning and appeared to allow "Normalization of Deviance" in the plant's process control limits and alarms.

- Knowledge and skills: Training and education for responding to upset situations and offsite response was inadequate.

- Process Monitoring and Control: Some of the MIC equipment was not functioning, was offline, or was inadequate to monitor the process and safety equipment properly. Compensating measures should have been considered to address the absence of key process safeguards.

---

The long-term consequences of "Bhopal" are still being felt, since the area has not been fully decontaminated. The Institution of Chemical Engineer's director of policy, Andy Furlong, stated in 2014:

*"... even though three decades have passed since Bhopal, we must never stop reminding ourselves that the lessons from the past are there to be learned, and crucially, acted upon."* (Furlong Press Release 2014)

The BP Texas City Example Incident 3.3 resulted in tremendous losses, in both the human casualties and injuries and financial losses.

## Example Incident 3.3 – BP Texas City 2005

The 2005 BP Texas City refinery explosion on the Isomerization Unit occurred during startup, when the Raffinate Splitter fractionation column overfilled (illustrated in Figure 3.1). The control board operator had failed to open the column bottoms line to tankage as required by the operating procedure. The bottoms temperature of the Raffinate Splitter exceeded the maximum temperature specified in the operating procedure. Although the column top pressure was normal, the hydrostatic head of liquid in the overhead line was sufficient to open the three relief valves at a lower level on the overhead line. Consequently, the liquid feed flowed directly to an atmospheric Blowdown Drum, allowing the liquid feed and vapors to release from the elevated vent stack on the Blowdown Drum and pool around the drum before finding an ignition source (CSB 2007).

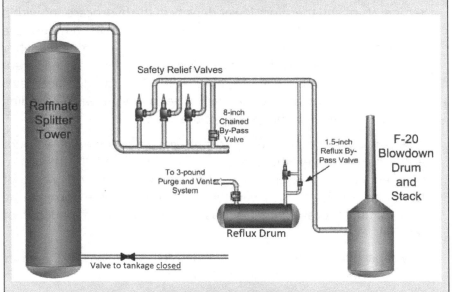

**Figure 3.1  BP Texas City Raffinate Splitter**

**Example Incident 3.3 – BP Texas City 2005 – (*cont.*)**

During the startup,the level indicated by the gauge in the Raffinate Splitter appeared to drop from 100% to 80% despite the feed rate remaining at 20,000 bpd (132 m$^3$/h) and a block valve remaining shut on the bottoms line. However, the actual level in the tower was far above the full-scale reading of the level gauge and increasing.

The level device and its associated transmitter were designed to measure the liquid level in a 5-foot (1.5 m) span such that 100% of its calibration corresponds to approximately 10 feet 3 inches (3.1 m) in a tower 164 feet (50 m) tall. Contrary to several published reports and technical papers, the displacer-type level device on the Raffinate Splitter worked as designed before, during, and even after the incident, when it was tested by an independent third-party expert.

The apparent reduction in level was a result of the higher temperature and therefore lower density of the hydrocarbons within the displacer level device, which did not have temperature compensation. As the bottoms temperature in the Raffinate Splitter increased, the density fell, which was reflected by the apparent reduction in indicated level from 100% to 80%. The displacer level device no longer measured the level in the column. It was responding instead to changes in the density of the fluid.

This confused the operator into believing that the level was not abnormal (in range) when, in fact, the column was full. One of two independent high-level alarms (the high-high alarm) was not functional, the high alarm was not acted upon, per the norm during startup, and a local sight glass was unreadable due to a buildup of residue.

The subsequent explosion and fires led to 15 fatalities, 180 injuries and financial losses exceeding $1.5 billion.

**Example Incident 3.3 – BP Texas City 2005 – (*cont.*)**

Lessons learned in relation to abnormal situation management:

- Operating procedures: This event happened during a transient operation when starting up the column. The startup procedure was not correctly followed, although the US Chemical Safety and Hazard Investigation Board (CSB) subsequently found that procedural deviations during startup had become common practice. (CSB 2007)

- Management of Change: Operating practices changed over time to reflect difficult control of startup, but management failed to address the safety implications of these changes and procedures were not updated.

- Process Monitoring and Control:

  o The process was allowed to operate outside normal limits for both the column bottom level and the temperature. Critical alarms were out of commission. A lack of situational awareness by the control panel operators resulted in no attempt to bring the column back into safe operating limits during the incident.

  o Providing suitable information and system interfaces for front line staff enables them to reliably detect, diagnose, and respond to potential incidents. The HMI should be designed so that the operator has multiple ways of viewing plant status, improving situational awareness that would make it easier to identify level gauge failure.

- Organizational roles and work processes: Personnel response to the incident was inadequate and lacked supervisory intervention. Shift handover documentation was insufficient to help incoming operators understand the gravity of the situation. Safety culture was poor.

- Work Environment: Inadequate definition of appropriate workload, staffing levels, and working conditions for front line personnel.

The investigation into the Texas City incident resulted in 81 recommendations and led to major changes in process safety management that went far beyond the immediate causes of the incident.

### 3.2.1    Process Control Systems –the First Line of Defense

The first line of automated defense is usually the process control system. These systems vary from simple analog or digital control instrument loops and alarms, through complex Distributed Control Systems (DCS), and even to Advanced Process Control Systems (APC). APC refers to advanced control techniques, such as feedforward, decoupling, multivariable, and inferential control that are often used to automatically optimize and monitor the operating variable set points to achieve production targets with lower costs, while adhering to the safe operating limits. CCPS has published two books that provide clear and concise information on guidance for control systems: *Guidelines for Safe Automation of Chemical Processes* (CCPS 2017c) and *Guidelines for Safe and Reliable Instrumented Protective Systems* (CCPS 2017b). When properly designed, tested, and maintained, these systems can provide a highly reliable approach to managing many abnormal process situations that were generally anticipated and had therefore been identified during the design process.

Some abnormal situations, however, may not be anticipated at the process design stage, when the protective safeguards are being developed, or when they are subsequently modified. These may include, but are not limited to, a liner that collapses in a vessel, agitators that decouple, a packing collapse in a column, a chemical composition change, a plugged instrument sensing line or nozzle, and in some cases, processes being impacted by certain weather extremes such as freezing of instrument lines or process vents. The ability to troubleshoot and quickly recognize such issues as abnormal situation-conditions is the goal of this book. Chapter 5 of this book discusses several tools and methods that fall under Hazard Identification and Risk Analysis (HIRA), including: Hazard and Operability Analysis (HAZOP), Fault Trees, and other methodologies that can be helpful in pre-emptively identifying abnormal situations at the design stage to ensure that adequate controls are provided.

When a control system is retrofitted to an existing process, it must undergo a full Management of Change (MOC) process including addressing how abnormal situations are handled and the additional training and competencies that would be required by the operating personnel. The CCPS book *Guidelines for Management of Change for Process Safety* (CCPS 2008c) references process control changes. It is highly recommended that representatives of operating staff be involved in the design and development of the interfaces and procedures associated with the new or modified control system, as they would be the frontline defenders against possible system failures and therefore must be familiar with how and why various controls are operated. Many incidents have occurred after a new control system was introduced with little or no previous input from operating personnel, resulting in difficulties for the operators during an emergency.

Example Incident 3.4 is an example of the consequences of introducing a new control procedure without first ensuring that operating personnel are familiar with the system.

### Example Incident 3.4 – Oklahoma Well Blowout 2018

In order to control an oil well that is being drilled, it is essential that the hydrostatic pressure of the drilling "mud" balances the pressure of the fluid in the formation. Normally this is achieved by the correct combination of static head and density of the mud. In the Oklahoma well blowout incident, when a drill pipe was being removed (a process known as "Tripping"), not enough mud was pumped down an underbalanced well to replace the displacement volume of the drill pipe. This led to the formation pressure exceeding the head of the mud, causing a blowout and fire that led to five fatalities.

Many factors contributed to this incident. One was the introduction of a new electronic version of the trip sheet, which could automatically calculate the fluid balance in the well, rather than relying on a manual calculation. However, the operator was not trained on how to use the electronic trip sheet. As a result, a significant gas influx built up in the well prior to the blowout. Further complicating matters, the drillers had turned off the alarm system that was giving excessive nuisance alarms, masking the more critical alarms (see also Example Incident 3.16).

The CSB report (CSB 2019) states:

> "Drillers are trained to monitor whether the well is taking the correct amount of mud while tripping by filling out a "trip sheet," where the calculated volume of the pipe removed is compared to the volume of mud taken by the wellbore. If the actual volume of mud pumped to keep the wellbore full is less than the calculated volume of the pipe being removed, it could be an indication of an influx of formation fluids into the wellbore."

**Example Incident 3.4 – Oklahoma Well Blowout 2018 – (*cont.*)**

The company had just started transitioning to using electronic trip sheets instead of paper trip sheets. The electronic trip sheet had built-in functionalities, including the ability to calculate parameters automatically. These were previously evaluated using the paper trip sheet. While one driller liked the electronic trip sheet because of the ease in populating fields, the driller on the shift during the incident indicated that he was not computer savvy and had only used the electronic trip sheets once before during an actual tripping operation.

The driller told the CSB he had never received formal training on using the electronic trip sheet, but tried to self-teach, using trial-and-error. This lack of training, in combination with the company not conducting performance assurance to determine if he could actually use the trip sheet in a practical situation, led to misusing the electronic trip sheet and contributed to the significant gas influx."

Lessons learned in relation to abnormal situation management:

- Knowledge and skills: The introduction of a new automated checklist system, with no adequate training or understanding of the new system on the part of the operator, was a key causal factor in this incident.

- Management of Change: The system must ensure that employees are involved in the details of procedural changes as well as process changes and are properly trained prior to operating the process.

- Process Monitoring and Control: The isolation of nuisance alarms must be carried out under a management procedure and "bad actors" should be dealt with as detailed in Section 5.3.2.

Whether long established or retrofitted, a well-developed and designed system should return the process to a safe operating condition, provide notification alarms to plant personnel that some action is needed, or initiate shutdown of the process to prevent an unwanted consequence.

### 3.2.2   Frontline Operators

History shows that it is often the initial lack of understanding, or an ineffective response by frontline personnel to alarm notification, that can allow an event to propagate and result in significant consequences. These issues can occur due to a number of factors:

- A poorly designed alarm management system, alarm flooding due to many simultaneous alarms (both genuine and sometimes erroneous), nuisance alarms, incorrectly prioritized alarms

- Inadequate or lack of personnel training, inexperienced personnel lacking knowledge or competence

- Lack of situational awareness, possibly due to a lack of training or experience

- Normalization of Deviance (accepting as normal the operation of a process outside the limits of established normal operation )

- Unauthorized or improper bypass of safety interlocks and alarms

- Inadequate staffing

- Human factors, leadership, and cultural issues, including "Stop Work Authority" where the operator is empowered to stop the process (see Chapter 2) and clarity of guidance is provided to operating personnel on when to call for outside help

Another frequent cause of abnormal situations is associated with instrument failure where there may be no effective alarms to alert the operator that something has gone wrong, as in Example Incident 3.5, the Buncefield explosion. For example, a level gauge may stick, and the level gradually changes outside the normal range, but the operator only becomes aware if a high or low level alarm activates - assuming, of course, that such an alarm has been fitted and is functional.

**Example Incident 3.5 – Buncefield Explosion, 2005**

Buncefield is a fuel storage terminal located about 3 miles (5 km) from the center of Hemel Hempstead, United Kingdom, and is connected by three pipelines to oil refineries, other fuel terminals, and airports. The site was permitted to store up to 194,000 metric tonnes of fuel, including aviation fuel (Jet A1), petrol (gasoline), diesel, and kerosene.

From Saturday, December 10 to Sunday, December 11, 2005, Tank 912 was being filled with gasoline from one of the pipelines. At about 5:20 am on Sunday morning, the tank began to overflow. Unobserved by the operators, the tank continued to overflow, forming a vapor cloud, until a huge explosion occurred at 6:01 am, followed by a series of explosions and fires that ignited 20 large fuel storage tanks nearby.

The explosion resulted when the vapor cloud ignited, possibly from the electrical switchgear associated with the firewater pumphouse. The vapor cloud was produced when over 250 m$^3$ (66,000 gallons US) of gasoline overflowed from the tank vents at the tank filling rate of above 550 m$^3$/h. (145,200 gallons/h). The magnitude of the explosion registered 2.4 on the Richter scale.

No fatalities resulted, although there were 43 non-serious injuries occurred. Properties nearby were destroyed, and damage extended to some 5 miles (8 km) away from the site. Environmental consequences from the smoke (visible from France), loss of containment from secondary containment (bunds) and the site effluent systems due to the quantity of firewater used and the type of foam. Total economic costs were estimated at £1 billion (IChemE Report 2008).

**Example Incident 3.5 – Buncefield Explosion, 2005 (*cont.*)**

The lighter fractions of the winter blend of gasoline had volatilized, forming a flammable cloud that was estimated to cover an area of some 80,000 $m^2$.

The immediate cause was the failure of the level gauge, which was stuck at a fixed reading, preventing the subsequent activation of a high-level alarm. An independent high-high level alarm also failed to activate. The operators had failed to observe or act on the stuck level gauge. They also did not estimate the time to fill the tank, instead relying on the instrumentation. The level gauge was known to be unreliable and the high-high level alarm, a float switch, incorporated a test level that had not been correctly repositioned or fitted with a padlock after a high-level trip test had been conducted. Locking of the level trip instrument was not fully understood by site personnel and the UK HSE issued an alert notice for users of similar equipment at other sites.

The Investigation Board report contained some 25 recommendations, several of which are relevant to a discussion on abnormal situations. Many of the recommendations were associated with improvements in the safety management systems, leadership, and culture on the site.

Lessons learned in relation to abnormal situation management:

- Procedures: Provide effective standardized procedures for key activities in maintenance, testing, and operations.

- Knowledge and Skills: Improve training, experience, and competence assurance of staff for safety-critical tasks and environmental protection activities.

- Work Environment: Define appropriate workload, staffing levels, and working conditions for frontline personnel.

- Communications: Ensure robust communications management between sites and contractors and with operators of distribution systems and transmitting sites.

---

**Example Incident 3.5 – Buncefield Explosion, 2005 (*cont.*)**

- Design: Other important considerations.

  o Consider installing gas detectors in bunds that have the potential for large quantities of flammable materials to be released.

  o Consider improvements to managing the integrity of overfill protection systems.

  o Ensure that the site receiving the fuel (rather than the transmitting location) has safe, ultimate control of tank filling and is not reliant on a third party to safely terminate or divert a transfer.

- Process Monitoring and Control:

  o Understand and define the role and responsibilities of the control room operators (including automated systems) in ensuring safe transfer processes. Improve procedures for product movements and record keeping.

  o Provide suitable information and system interfaces for frontline staff to enable them to reliably detect, diagnose, and respond to potential incidents. The HMI should be designed so that the operator has multiple ways of viewing a transfer status, improving situational awareness that would make it easier to identify level gauge failure.

---

One of the outcomes of the Buncefield incident was an update to the API 2350 standard and the application and use of Automatic Overfill Protection Systems (AOPS) for high-risk scenarios.

Some principles for the effective management of abnormal situations are included in several elements of the CCPS Risk Based Process Safety framework: Hazard Identification and Risk Analysis (HIRA); Operating Procedures, including safe operating limits, and consequences of deviation from safe limits; Training and Performance Assurance; Asset Integrity and Reliability; Conduct of Operations and

Operational Discipline; and Process Safety Culture. These are discussed in more detail in Section 3.3.

The next section provides further details on some of the factors that can have an impact on the management of abnormal situations.

## 3.3    MANAGEMENT OF ABNORMAL SITUATIONS AND LINKS TO RISK BASED PROCESS SAFETY

Focus areas detailed under Section 3.1, comprising the ASM® Consortium research topics and additional areas, all encompass various elements of Risk Based Process Safety (RBPS) as detailed in the CCPS book, *Guidelines for Risk Based Process Safety* (CCPS 2007a).

Table 3.1 lists the RBPS elements and how they relate to prevention of process safety incidents.

The 20 elements from the CCPS guideline book are associated with four main RBPS accident prevention pillars:

- Commit to Process Safety,

- Understand Hazards and Risk,

- Manage Risk, and

- Learn from Experience.

## Table 3.1  Process Safety Accident Prevention Pillars and RBPS Elements

| Process Safety Accident Prevention Pillars and Focal Points | RBPS Elements |
|---|---|
| *Commit to Process Safety* | |
| Ensure management cares and provides adequate resources and proper environment<br>Ensure employees care<br>Demonstrate commitment to stakeholders | Process Safety Culture<br>Compliance with Standards<br>Process Safety Competency<br>Workforce Involvement<br>Stakeholder Outreach |
| *Understand Hazards and Risk* | |
| Know what you operate<br>Identify means to reduce or eliminate hazards<br>Identify means to reduce risk<br>Understand residual risk | Process Knowledge Management<br>Hazard Identification and Risk Analysis |
| *Manage Risk* | |
| Know how to operate processes<br>Know how to maintain processes<br>Control changes to processes<br>Prepare for, respond to, and manage incidents | Operating Procedures<br>Safe Work Practices<br>Asset Integrity and Reliability<br>Contractor Management<br>Training and Performance Assurance<br>Management of Change<br>Operational Readiness<br>Conduct of Operations<br>Emergency Management |
| *Learn from Experience* | |
| Monitor and act on internal sources of information<br>Monitor external sources of information | Incident Investigation<br>Measurement and Metrics<br>Auditing<br>Management Review and Continuous Improvement |

### 3.3.1   Commitment to Process Safety

Commitment to process safety and prevention of process safety incidents begins with the management leadership team and their establishment of a *Process Safety Culture* that supports risk identification and safe work practices. Establishing organization roles and *Workforce Involvement* along with shared responsibilities sets the foundation for identifying and preventing abnormal situations. For example, the leadership team will need to provide expert resources, sufficient time, and the required funds for the necessary risk assessments to be appropriately completed otherwise an abnormal situation could result in a significant unanticipated hazardous event with serious consequences.

### 3.3.2   Understand Hazards and Risk

The RBPS elements associated with the understanding of abnormal situations are contingent on identifying the risks with a *Hazard Identification and Risk Analysis (HIRA)* study conducted when the system is designed or when modifications are made. Tools that may help anticipate potential abnormal situations during these phases are included in Chapter 5. Where the hazards have been identified, these are then incorporated into the RBPS elements *Process Knowledge Management* and *Operating Procedures*, (Section 3.3.3) and to improve understanding of the procedures, *Workforce Involvement* (Section 3.3.1) is a key element to ensure the accuracy, practicality, and relevance of the procedures. The element *Workforce Involvement* further requires developing and communicating a written plan of action regarding worker participation that should include management of abnormal situations.

### 3.3.3   Manage Risk

Safely managing expected or unexpected events is a key responsibility of frontline personnel. If a process has been designed following RAGAGEP and specific industry standards, the likelihood of unwanted events progressing to a serious consequence is greatly reduced. However, there may be those abnormal situations that were not anticipated or included in the safe design of the process. These possible events can best be identified and safely managed if the frontline operating team has been involved in the HIRA of the process; participates in writing, reviewing, and training on *Operating Procedures* and *Safe Work Practices*; and is included in *Operational Readiness* to start up or re-start a process following either a scheduled or unexpected shutdown. *Training and Performance Assurance* of operation personnel to reliably perform during normal operations as well as during upset or abnormal process situations is vital to ensuring that a situation does not escalate to a serious event. Section 3.4 covers procedures and training in more detail.

### 3.3.4    Learn from Experience

Another key to managing abnormal situations is *Learning from Experience;* one of the RBPS pillars. It is far less costly to benefit from the learning of others rather than to repeat the experience. Many serious events and their cause(s) are in the public domain and by careful analysis, much of the learning, both from incidents and near-misses, can be applied to other facilities. This includes reviews of the learning from historic events on the same or similar processes that may well still apply today. Where it is relevant to individual processes, this learning should be institutionalized via mechanisms such as procedures, policies, or corporate standards. History has shown that sometimes, during an *Incident Investigation*, senior front-line personnel will share that a similar event happened years before. Therefore, understanding and sharing the learnings from internal as well as external events can help to prevent abnormal situations or events occurring in the future. The CCPS Process Safety Incident Database (CCPS/PSID) is a resource for learning from case histories. The CCPS Process Safety Beacon (CCPS/*Process Safety Beacon*) discusses many different causes for process incidents and shares actual incident information.

*Auditing* process safety is another way of identifying and correcting management system issues and weaknesses before they result in abnormal situations and potentially serious incidents. Both internal/self-audits and external audits conducted periodically, lead to actions, to improve the effectiveness of the management system, with an emphasis on abnormal situations.

### 3.3.5  Additional RBPS Elements Related to Management of Abnormal Situations

The RBPS elements in this section are briefly discussed in this chapter for awareness. Some of them, such as MOC, will be discussed in Chapter 5.

#### 3.3.5.1  *Compliance with Standards*

Includes applicable regulations, standards, codes, and other requirements issued by national, state/provincial, and local governments, consensus standards organizations, and the corporation. Interpretation and implementation of these requirements include development activities for corporate, consensus, and governmental standards.

#### 3.3.5.2  *Process Safety Competency*

Addresses skills and resources that the company needs to have in the right places to manage its process hazards. Provides verification that the company collectively has these skills and resources and applies this information in succession planning and management of organizational change.

#### 3.3.5.3  *Asset Integrity and Reliability*

Activities to ensure that important equipment remains suitable for its intended purpose throughout its service. Includes proper selection of materials of construction; inspection, testing, and preventive maintenance; and design for maintainability.

#### 3.3.5.4  *Management of Change*

Process of reviewing and authorizing proposed changes to facility design, operations, organization, or activities prior to implementing them, and updating the process safety information accordingly.

#### 3.3.5.5  *Conduct of Operations*

Means by which management and operational tasks required for process safety are carried out in a deliberate, faithful, and structured manner. Managers ensure workers carry out the required tasks and prevent deviations from expected performance:

### 3.3.5.6   *Measurement and Metrics*

Leading and lagging indicators of process safety performance, including incident and near-miss rates as well as metrics that show how well key process safety elements are being performed. This information is used to drive improvement in Process Safety. Safe Operating Limit Excursions and Demand on Safety Systems are two example metrics that are relevant to abnormal situations. Metrics will be further discussed in Chapters 5 and 6.

### 3.3.5.7   *Management Review and Continuous Improvement*

The practice of managers at all levels of setting process safety expectations and goals with their staff and reviewing performance and progress towards those goals. This may take place in a staff or "leadership team" meeting or individually. The practice may be facilitated by the process safety lead but is owned by the line manager.

## 3.4   PROCEDURES AND OPERATING MODES FOR MANAGING ABNORMAL SITUATIONS

This section addresses how to write and structure procedures that incorporate principles describing how to manage abnormal situations that can then be used by operating personnel to make appropriate decisions during periods of abnormal situations. However, many abnormal situations will not be anticipated, and for those, a more holistic approach to abnormal situation management is required, which includes not only written procedures, but also training personnel to recognize an abnormal situation, protocols for dealing with it, and providing the resources to respond to events that may not be foreseen.

### 3.4.1   General Principles for Procedure Development

It is outside the scope of this book to provide detailed guidance, or templates, for development of normal operating procedures, however, these can be found in other references including *Guidelines for Writing Effective Operating and Maintenance Procedures* (CCPS 1996). However, for managing and controlling abnormal situations, some high-level human behavior principles apply, as summarized in the sources discussed in this section.

Many studies of human behavior (Swain & Guttmann 1983, Embrey et al 1984, CCPS 2004, Wincek & Haight 2007) indicate that the greater the seriousness of a situation, and the less time that is available to respond, the higher the probability that an error or incorrect response will result. Human error, and the psychology behind it, is also discussed in detail in James Reason's book (Reason 1990) that includes details of Rasmussen's skill-rule-knowledge classification of human performance and the generic error-modelling system (GEMS), within which to locate the origins of the basic human error types (Rasmussen 1982). Further details are beyond the scope of this text, but the references are provided.

Therefore, it is important to develop emergency response procedures and capabilities that are suited to the criticality of the situation – that is, the more serious the situation, the more prescriptive, straightforward, and simplified the guidance should be. This concept is illustrated throughout the rest of this section.

Once a procedure is developed, it can be critically reviewed using an informal process such as circulating the draft to several future users of the procedure, or a more formal process such as "Procedure HAZOP" can be utilized. In either case, a simple but useful self-review approach to a procedure involves walking through the procedure and for each step asking these questions:

- Is the step accurate? [e.g., is the temperature supposed to be raised to 250 degrees, not 200 degrees?]

- Is the step complete? [e.g., should the operator wear "neoprene boots" rather than just any "boots"?]

- Is the step concise? [or Is the step so long that the operator can get confused and leave something out?]

- Is the step in the right sequence? [e.g., should Step 5 come before Step 3?]

Depending on the criticality of the step, a fifth consideration can be added to the list:

- What are the consequences of failing to follow the step?

In circumstances where the correct use of a procedure by front-line personnel is critical to prevent an abnormal situation from proceeding to a significant event, an evaluation of the likelihood of the procedure being completed correctly is recommended. This assessment can be completed by an analysis method referred to as Human Reliability Analysis (HRA), as described in the CCPS book, *Guidelines for Hazard Evaluation Procedures*, (CCPS 2008b). HRA is a method to evaluate whether necessary human actions, tasks, or jobs will be completed successfully within the required timeframe. Additionally, the HRA can be used to determine the probability that no extraneous human actions detrimental to the system will be performed. The HRA results can then be combined with the potential impact of the significant event impact to determine the scenario risk and evaluate whether existing safeguards are adequate.

For abnormal situations that are anticipated, a procedure can be in the form of a written checklist, or possibly an automated checklist that links to the DCS that controls the process. These are typically developed for situations such as loss of services (steam, power, air, inert gas, or other utilities) but they may also include bomb threats, logistical issues, road closures, chemical spills, or incidents from outside the site.

Automated checklists, which pull data from a system and provide operators with a summary of the checks, must be developed very carefully and must properly consider human factors. A paper (Mosier et al [1992 Paper] 2016) involved a trial in the aviation sector on the use of electronic checklists versus manual checklists. Mosier concluded that:

> *"Making checklist procedures more automatic, either by asking crews to rely on system state as indicated by the checklist, rather than as indicated by the system itself, will discourage information gathering and may lead to dangerous operational errors."*

For the process industries, the best way to gather information is to go to the original source of the data rather than relying on a system's automated interpretation of its status. Checklists are a useful tool in encouraging performing tasks correctly, but where these have been automated, the user should also use other available information to verify that the checklist actions have been completed correctly.

Abnormal situations that are not anticipated provide a much greater challenge in terms of procedures. For example, an instrument failure, such as a pressure sensor on a pressure control loop getting stuck, may result in an overpressurization of a tank and a relief valve opening. In turn, this may lead to an incorrect action such as an operator isolating the relief system, instead of validating the pressure sensor performance with a local pressure gauge. Faults with instrumentation may also result in the incorrect action of more sophisticated control systems (common mode failure), providing an even greater challenge. Similar challenges with the effectiveness of procedures are associated with the more complex systems that are becoming more prevalent with advances in technology. Competencies and skills on the diagnosis/troubleshooting of potential faults with instrumentation should form part of a generic set of procedures that are regularly exercised. Training and exercise may be part of a computer simulator or a simpler, manual process following a chain of logic. Further details on education and training are provided in Chapter 4.

Local sensitivity to environmental (or visually noticeable) releases, and cultural factors, may also need to be considered as illustrated in Example Incident 3.6.

**Example Incident 3.6 – Relief Valve Opening**

While returning from lunch one day, a unit engineer noticed a trail of condensate running in a line along the ground under the pipe rack. The engineer knew from experience that such a trail is frequently a sign that a light material such as LPG is vaporizing in a pipe (in this case a flare header), which cooled the pipe and caused atmospheric moisture to condense. As he followed the trail back to the source (an LPG treater), he noticed an operator at the top of a vessel actively blocking in a relief valve. When asked, the operator replied that he was blocking in the relief "because my supervisor said it is making the flare too large."

The unit engineer stopped him before a disaster occurred, but this illustrates those issues such as culture and hypersensitivity to events that are visible to outsiders can motivate workers to take inappropriate and potentially dangerous actions.

Lessons learned in relation to abnormal situation management:

- Understanding abnormal situations/process safety competency: In this case, the operator (and his supervisor) failed to consider the potential consequence of blocking in the relief valve.

- Management of Change: Temporary or permanent changes to any protective device should be evaluated through the MOC process.

- Procedures: The procedures (and associated training) for diagnosing a faulty relief valve should include requirements to:

  o Check on other pressure gauges in the system, accounting for difference in head (height of fluid)

  o Replace the pressure gauge to ensure it is working correctly

  o Check pressure control system for correct functioning

  o Obtain senior management approval before isolating any relief system, via structured and formal temporary bypass procedure

### 3.4.2   Operating Modes

Abnormal situations can occur during different operating modes. Studies completed by CCPS have shown that process safety incidents occur five times more often during startup than during normal operations (CCPS 1995b/citing Marsh McLennan 1992). Similarly, 50% of incidents in the refining industry occur during startups, shutdown, and other events that infrequently occur (Ostrowski & Keim 2010).

This section includes examples for each of the following operating topics:

- Operating Modes
    - o   Startup/commissioning
    - o   Normal operations
    - o   Emergency operations
    - o   Shutdown/decommissioning
    - o   Change between operating modes and products/grades
    - o   Type of operation:  Include guidance on the impact of operating type (batch, semi-batch, continuous, transient operation) as it relates to abnormal situations.
- Types of Material Being Processed
    - o   Flammability – flammable or combustible
    - o   Phase – solid, liquid, vapor, mixed
    - o   Other hazards including – toxicity and chemical reactivity

### 3.4.2.1   *Startup/Commissioning*

At no other time is there greater opportunity for an abnormal situation than during initial commissioning of a new unit. The equipment and perhaps the process itself is somewhat unfamiliar to the operating staff, and the subtle effects of design implementation details have not yet become apparent.

Furthermore, there are often many different contractors conducting their individual specialized tasks, such as the piping portion of a project, or just the engineering. Communication is critical to ensure no important information is missed when the project is passed from the design engineers to the production team.

For this reason, startup and commissioning procedures are usually very prescribed, and presented in a lengthy but systematic format. Most of the process of writing startup/commissioning procedures should follow standard practices, e.g.:

- Identify the individual steps (actions) to be taken.

- Provide the milestones (e.g., reactor temperature) that are to be achieved before moving on to the next step.

- Identify potential dangers associated with each step.

Example Incidents 3.7 and 3.8 provide examples of problems that can manifest themselves during initial commissioning. In the first example incident, the event was anticipated; in the second, it was not.

---

### Example Incident 3.7 – Hyperactive Catalyst Runaway

In many processes involving hydrodesulfurization of the feedstock, the reaction catalyst is provided in a hyperactive state that can frequently lead to a runaway condition when the catalyst is first exposed to the feed. For this reason, it is typical to have a 'presulfiding' step where a baseline quantity of sulfur is deposited on the catalyst to reduce its activity. Even so, temperature excursions do occur during the initial introduction of feed. The procedure for addressing the excursion is generally straightforward and governed as an insert or attachment to the startup procedure.

Lessons learned in relation to abnormal situation management:

- Knowledge and Skill: Knowledge with respect to unique catalyst characteristics is needed to prevent unwanted consequences.

- Procedures: Procedures with special warnings can be a safeguard to prevent excursions.

### Example Incident 3.8 – Distillation Column Startup

In one process, a distillation column was used to separate components of a reactor effluent stream that ranged from heavy oils to propane-range material. During commissioning, the contractor did a very poor job of removing scale from the tower, and when the tower was started up, all the debris flowed to the bottom of the tower and collected in the bottoms pumps suction strainers. The bottoms pump lost suction pressure and was not able to pump out the desired flow rate. The backup pump was started, but its suction strainers also quickly plugged.

The strainer for the primary pump required removal for cleaning, but it was not able to cool below the process material's autoignition temperature before the backup pump's strainer also plugged. The response was to shut down a plugged pump, pull out its strainer while applying a steam hose to the strainer (to reduce the local oxygen concentration and prevent auto-ignition), clean the strainer and then restart the pump. All this was done while operating in a restricted manner using the spare pump, during the short time that it could operate before its strainer also plugged.

Typically, a "procedure" would not be created for this unexpected activity, but this case does illustrate the need for operators and supervisors to have guidelines on how far they are authorized to institute a hazardous activity before making a decision to shut down the unit.

Lessons learned in relation to abnormal situation management:

- Understanding abnormal situations: Fully understanding a situation and potential consequences is needed to avoid making high-risk decisions rather than shutting down a process.

- Communications: A discussion between contractors, operators and plant engineers of the plugging issue should have occurred before initiating any actions to address the situation.

### 3.4.2.2 Normal Operations

Abnormal situations can also arise during normal operations. Sometimes, these can be anticipated and provided for in advance through specific procedures – for example, in response to power failures or other loss of utilities. Other common situations include a pump trip or a control failure.

The provision of sufficient, appropriately trained operators and field personnel with adequate resources and good communications should ensure an appropriate response to these types of upsets. Training could be provided using process simulators, similar in concept to that provided to airline pilots. Some upsets may not be anticipated, however, as detailed in Example Incidents 3.9 and 3.10.

## Example Incident 3.9 – Hydrocracker Operations

In an oil refinery hydrocracking unit, a feed pump failure had historically been addressed by simply starting the spare pump, then addressing any temperature excursions that might occur during the event in the customary manner. However, a modification was made in which the upper heater chamber was combined with the chamber from another heater. The redesign was not approved by the licensor. The effect of the change was to allow heat to flow from one side of the heater to the other, so that in a trip situation, heat continued to flow where it was not required. The first time the feed pump tripped, and the standard pump trip response was performed, the result was an extreme reactor runaway condition (from 400 °C to 1000 °C). This showed the need for an all-purpose emergency shutdown procedure that could be applied under any circumstance.

Lessons learned in relation to abnormal situation management:

- Management of Change: Implementing a change without considering input from the licensor on the potential effects of the change.

- Root Cause Analysis:  Less severe versions of this event had occurred at other similar units at other locations having the same design change in the preceding months. An analysis of the failure causes and an adjustment to the default response would have mitigated the damage that resulted.

Procedures: The procedure for starting the spare pump was simply based on historical experience with a more conventional heater design. This should have been re-examined as part of the Management of Change process.

---

### Example Incident 3.10 – Tower Flooding

Poor separation in a distillation column can be the result of insufficient flows of vapor/liquid due to low reflux and/or low boil-up from the reboiler.

In one typical case, the tower was providing poor separation of the individual components. The operator responded by adding more heat to the reboiler to provide more energy for the separation. This increased the tower overhead temperature/flow and a corresponding high level was reached in the overhead receiver, which was then addressed by adding more reflux. However, the separation of components in the tower worsened, and so the control room operator repeated the more-reboiler/more-reflux cycle over a period of an hour, with progressively worse results.

This was a case of tower 'flooding' in which liquid and/or vapor rates are too high. This leads to excessive liquid on individual trays, resulting in poor liquid-vapor disengagement, high pressure drop, and poor overall separation of components.

The solution was to remove the heat source to the tower, let everything fall to the bottom, and start it back up again.

Lessons learned in relation to abnormal situation management:

- Understanding Abnormal Situations: This is less about having a procedure to address the issue, and more about abnormal situation identification and training in the principles of distillation column operation.

Knowledge and Skill Development: Additional knowledge and training may have prevented the lead-up to the flooding. Learning from others—experience is a valuable knowledge-sharing tool.

---

Cultural influences may play a factor in choosing the interface between the control panel operator and the control panel, in order to allow more rapid detection of an abnormal situation. The following discussion between a licensor's representative and the control panel operators on a new unit illustrates this concept in Example Incident 3.11.

**Example Incident 3.11 – Control Panel Differences on Two Similar Units**

A facility had a hydrocarbon cracking unit that had been running for decades, built in an era when analog indications (temperature dials, strip charts, etc.) were industry standard. During a plant expansion, a parallel cracking unit was installed with more modern distributed controls and computer screen monitors (HMIs).

The unit engineer decided to ask the night shift control panel operators for their opinion on the 'fancy' control panel on the new unit compared to the 'dials and charts' panel on the existing unit. Their stunning response was, "We don't like the new system as much because it takes more time to figure out the condition of the unit *when we wake up*." In this facility, it had been customary for the control panel operators to take a nap during the night shift, and upon waking, be able to check two or three key dials on the control panel to see how things were going. In the 'new and improved' system, the control panel operators were obliged to scroll through a series of computer screens for the same information.

Lessons learned in relation to abnormal situation management:

- Organizational Roles and Responsibilities: The control panel operators should not be sleeping at all while on shift, which must be addressed through leadership and culture.

- Work Environment: This example incident description does not discuss the control room work environment, although if control room conditions could contribute to napping (e.g., poor lighting or ventilation), those conditions should be carefully analyzed and mitigated.

- Process Monitoring and Control: The situation does highlight possible limitations in the design of the HMI that should provide a single overview screen, enabling operators to rapidly assess the key process parameters that may be helpful in the event of an abnormal situation. Clearly, the HMI was not designed to maximize situational awareness when a rapid assessment of operating status was required.

### 3.4.2.3 *Emergency Operations*

Facilities should have operating procedures for emergencies that are anticipated and predictable, usually via the risk assessment process. Such situations might include loss of services such as:

- Electrical power
- Steam
- Process air
- Instrument air
- Inert gas
- Cooling medium (Cooling water, Chilled water, refrigerant)

Other emergency operating procedures might include provision for:

- Bad weather (storm, hurricane, ice)
- Raw material or catalyst shortage
- Full product stock tanks

However, not all such scenarios can be predicted, and it is important that operating staff recognize when such events are imminent or are occurring. One example would be a change in business conditions leading to a situation where a process is initially shut down temporarily, but the shutdown period is then extended until profitability returns.

Management and operating personnel must recognize that such an operating mode is outside the norm (i.e., an "emergency") and act appropriately to ensure that ongoing risks continue to be controlled. This might involve an emergency risk assessment being conducted by appropriate key personnel, which would be reviewed as the situation developed. This is particularly true for an event that has potential safety or environmental implications beyond the boundaries of the unit – for example, a major fire or toxic release. In these situations, a timely and correct response may be the difference between a relatively minor monetary loss or a major event that shuts down the facility or causes loss of life.

This type of situation is one in which the stress and time-criticality of the event is more likely to elicit an incorrect response. Where possible, the response should be dictated by a procedure that is simplified and prescriptive as much as possible. However, it should be noted that the development of and training for such procedures is difficult to do and often includes mostly table-top/what-if drills, which do not properly mimic mindset stress.

In cases requiring notification of staff outside the unit, or public responders, it may be useful to develop a hand-held flip chart that directs the supervisor regarding the order of actions and notifications, provides contact numbers and other critical information needed for the response. While the intent is not to eliminate judgment and initiative from the response, the flip chart reduces the number of decisions the responder needs to make, thus making the responses more likely to be correct.

Example Incident 3.12 and Example Incident 3.13 include a loss of power event and a serious incident following an extended plant outage.

---

### Example Incident 3.12 – Loss of Site Power Supply

The electrical power supply to an ethylene cracking plant failed, resulting in a cascade of failure of site services including steam and instrument air. The Uninterruptible Power Supply (UPS) to the DCS failed after a matter of minutes, and lighting to the control room, which was designed to be explosion-resistant and therefore not fitted with windows, was lost, plunging the control room into darkness.

The hydrocarbon feeds to the cracker tripped as designed in the emergency response system. Despite this trip, the cracker tubes ruptured due to thermal stress from a loss of temperature control. The loss of temperature control resulted from a loss of steam flow that should have allowed for controlled cooling. Cracked gas flowed back into the crackers from the downstream quench system and ignited in several of the crackers, causing fires and further damage.

The site had identified that loss of power to the site was a key safety consideration. The plant had two power feeds that they considered independent, and the site could operate on just one supply. However, they both failed due to a fault at a common substation owned by the power company.

**Example Incident 3.12 – Loss of Site Power Supply (*cont.*)**

The site had identified critical power supplies and provided UPSs and several self-starting backup generators for when the power supplies failed. The generators were tested on a monthly basis, but the test did not include synchronization to the site systems. When they were required, several of them failed to synchronize and tripped out, resulting in loss of critical power supplies to the control system for a high-integrity steam generator. Once that was lost, steam supplies to critical turbine-driven machinery failed, leading to loss of compressed air for the instrumentation system.

Each of the outlet valves from the 12 crackers was motor-driven with a manual handle in case of failure of the motor drive, as was the situation here, although each one required over 100 turns before the valve could be closed. Operators started to close these valves, once they realized that the cracker tubing was rupturing, to prevent the reverse flow of hydrocarbons. However, there was not enough time, or enough staff, to close all 12 valves before the major, uncontrolled fires were burning in the crackers.

Lessons learned in relation to abnormal situation management:

- Services Failure: The site thought that they had adequately prepared for this type of situation via the HIRA process, however, the backup systems failed when they were required.

- Procedures: There were procedures in place to test backup utility services, although these did not fully simulate emergency conditions of total power loss. The backup to the backup (manual closure of valves) was impractical due to lack of time and labor availability.

---

**Example Incident 3.12 – Loss of Site Power Supply (*cont.*)**

- Work Environment: Operators were literally working in the dark. Backup power for control systems and critical lighting must be very reliable and last for the required time. That time must be assessed by detailed analysis and supported by scenario-based exercises (drills.).

- Abnormal situation recognition: The facility and plant personnel had considered normal upsets and concluded that the plant had adequate backup safeguards. The abnormal situation of a common fault failure of offsite substation electrical supply as well as failure of the emergency power generators to synchronize was not considered. The risk analysis may need to review reliability of critical services that are located beyond the fence-line.

---

Failure of critical services on a high-hazard facility requires robust backup systems that are properly tested to ensure they will work when required. The failure of backups is a common situation since the systems are often designed without considering their testing. It is often not possible to test such equipment without risking tripping out some or all of the plant.

Third party services that ensure that the facility operates safely should be considered during risk analysis. These include electrical supply and inerting gas such as nitrogen. These utilities are often contracted from suppliers that may not understand the risks posed to an operating plant if supplies are interrupted. Backup plans for loss of supply should be discussed and arranged in advance with the suppliers.

Example Incident 3.13 discusses a failed restart of a chemical factory following a prolonged outage, resulting in multiple fatalities (Hailwood 2020).

**Example Incident 3.13 – Styrene Runaway Reaction and Release, 2020**

On 7 May 2020, there was a leak of gas from a styrene storage tank at a chemical factory in Visakhapatnam, Andhra Pradesh, India, causing at least 12 fatalities, affecting some 3000 people and leading to an evacuation of a 2 km (1 ¼ mile) radius around this site.

The plant was in the process of restarting, following a temporary shut down due to lockdown imposed by the COVID-19 pandemic. The incident was a result of an uncontrolled exothermic self-polymerization reaction of styrene and a release of an estimated 800 metric tonnes of styrene gas from a storage tank.

Styrene monomer is a raw material used for the manufacture of plastics. It is a liquid at room temperature and has a boiling point of 145 °C (293 °F). Exothermic polymerization of styrene in storage is a known hazard, as a runaway reaction can occur once the temperature exceeds about 65 °C (149 °F), which leads to vaporization and release of styrene gas from the storage tank vents/relief devices. Normally, an inhibitor such as 4-tert-butylcatechol (TBC) is added as a stabilizer and polymerization inhibitor, but it is important to monitor the TBC concentration as it decreases with time. TBC requires the presence of oxygen to function and becomes less effective if the concentration of oxygen is lower than about 3% in the headspace, or if the material is stagnant (Plastics Europe 2018). The recommended optimum oxygen level in the headspace is 5% by volume; above 8% can lead to a flammable mixture. The recommended storage temperature should not exceed 25 °C (77 °F).

Under normal operation, the movement of styrene and its replenishment with fresh, stabilized material may have been sufficient to maintain its stability. Furthermore, operators have been monitoring the tank temperature and TBC levels and topping up with TBC if required. All of these factors were reduced or absent given the shutdown and reduced staffing; additionally, it was common practice to shut down the tank refrigeration system overnight.

**Example Incident 3.13 – Styrene Runaway Reaction and Release, 2020 *(cont.)***

The only temperature indication in the storage tank concerned was at the base, and the pressure safety valve from the tank emitted directly to the atmosphere and not via a flare system or other emission control system.

Lessons learned in relation to abnormal situation management:

- Abnormal Situation Recognition: The temporary closure of a facility requires special consideration, particularly if the duration of the closure is unknown at the start.

- Organizational Roles: Responsibilities and Work Processes: With a reduced workforce, revised procedures and responsibilities should have included extra checks on high-risk areas. These considerations should fall under a Management of Operational Change (MOOC) procedure.

- Procedures:  Procedures should have been revised to address high-risk scenarios due to work force and organizational changes.

- Process Monitoring and Control: Provided the risk had been identified, additional automatic control measures, including better temperature measurement could have been specified.

### 3.4.2.4    *Turnaround/Shutdown/Decommissioning*

Turnaround, shutdown, and/or decommissioning of process plant equipment is a normal activity during the life cycle of a chemical process. Because of this both scheduled, preventive, and unplanned emergency maintenance of mechanical equipment has become a critical function of highly performing organizations. Repair costs, production downtime, leak containment integrity, and regulatory inspections are all factors involved in developing and executing maintenance shutdown strategy. Advanced mechanical integrity planning methods include sophisticated tools such as Risk Based Inspections (RBI), Reliability Centered Maintenance (RCM), thickness testing, and monitoring. Asset Integrity engineers and maintenance specialists who have received specific training will typically have accountability for leading and performing maintenance of the equipment. Standardized mechanical integrity programs are now routinely utilized across many industries.

For example, Asset Integrity Management (AIM) is the recognized generic name for a process safety management system for ensuring integrity of assets throughout their life cycle. AIM is considered of such high interest and importance that CCPS has published a book on *Guidelines for Asset Integrity Management* (CCPS 2017a).

Written procedures for shutting down the process and associated equipment are necessary; otherwise, loss of chemical containment may occur, equipment can be damaged, people may be injured, and delays in maintenance timing and execution may affect plant production. Therefore, time-proven shutdown steps, associated preparation activities, and required safety measures are recommended to be in place with personnel training completed. During shutdowns even with the best planning, unknowns can occur, such as valves not closing fully, pipe and equipment drains experiencing plugging, solids being present in the bottom of vessels, knockout systems being temporarily out of service, and finding damaged internal equipment. Although these types of unknowns may not be considered abnormal situations, planning and discussing these potential unknowns in advance with plant operators and maintenance team members may prevent a situation from resulting in an unwanted event.

Decommissioning can be an AIM concern and could become an abnormal situation if re-use of the asset or assets is not expected but due to changing business requirements it is decided to return the asset to service. For example, after equipment is taken out of service, the asset may be "mothballed", and then stored in a spares compound on site. However, the business may envision an opportunity to save money and decide to return the asset to service. Alternatively, the entire process might be decommissioned and not expected to return to service, but the business subsequently says that demand for the product has returned. Plant operation and maintenance personnel are then challenged to return the process to service. Therefore, it may be forthright for decommissioning procedures to include the unplanned situation of having to return the process to service, as described in Example Incident 3.14. This could even be after years of being out of service. Alternatively, the MOC process may be utilized as the initial step to return decommissioned equipment to service.

---

### Example Incident 3.14 – Reboiler Decommissioning

A distillation column had two reboilers; one was relatively new and the other near end-of-life. The column only required one reboiler to be on-line. The decision was made to decommission the older reboiler as it had some steam leaks and a future capital investment (CAPEX) plan included provision for its replacement.

A detailed decommissioning procedure was written and approved including some specific inspection requirements to ensure that the reboiler was properly cleaned. The reboiler was then safely disconnected from the column and blind flanges were installed on the column nozzles. The reboiler tubes were to be water flushed to remove any residual chemical traces and solids. The tubes were then to be dried and checked for moisture. In addition, a physical inspection was to be conducted on each tube to ensure they were clean. The responsible personnel checked for moisture as required and, finding none, concluded that the reboiler was dry and the physical inspection of each tube was not necessary. This fact was not shared with the plant management.

Six months later, the newer reboiler experienced a mechanical problem and since the new CAPEX reboiler had not yet been received, the decision was to return the removed reboiler to service. However, the old reboiler failed its pressure test and several tubes were corroded and had some small holes. The investigation found that a small amount of solids had remained in some of the tubes and reacted with water to form acid. Over the six-month period, the acidic solids had created pinhole leaks. The personnel responsible for the inspection admitted to not conducting an inspection of all the tubes and stated that the reboiler tested dry and an inspection of a few tubes did not show any solid residue.

---

---

**Example Incident 3.14 – Reboiler Decommissioning (*cont.*)**

Lessons learned in relation to abnormal situation management:

- Procedures: The conclusion of this event is that despite a very detailed decommissioning procedure with clear instructions, it was not fully followed.

- Understanding abnormal situations: deciding to shortcut the written procedure with the potential consequence not understood.

- Communications: The responsible personnel checked for moisture as required and, finding none, concluded that the reboiler was dry and the physical inspection of each tube was not necessary. This fact was not shared with the plant management.

---

### 3.4.2.5   *Change in Operating Modes or Products/Grades*

Some operations use the same equipment in different modes, or for processing different products. Examples include:

- A batch reaction that proceeds under pressure, but at the end of the cycle is switched to vacuum mode to strip away the solvent used in the reactor.

- One product type is replaced with another, but the same equipment is used for each successive product campaign.

These multi-functional operations introduce the potential for several hazard situations, including:

- Introducing air, or passing through the flammable range of a material, during transition from positive to negative pressure.

- Product contamination or incompatible chemical mixing.

- Introduction of outside contaminants especially in processes that frequently switch over products and grades.

- Operating envelopes and limits may be different, e.g., flammable limits, thermal stability.

- Process equipment and controls may require different description, e.g., on DCS.

- Instrumentation may require different calibration to account for differences in physical properties such as density (See Example Incident 3.3, Texas City)

A thorough risk-assessment process, such as a "What-If" PHA should identify these issues, and adequate procedures/ checklists must be put in place.

Example Incident 3.15 and Example Incident 3.16 illustrate what can go wrong when operating procedures and design alarm issues are not aligned.

---

### Example Incident 3.15 – Batch Reaction Alarms Ignored

A polymer plant comprised 12 batch reactors, each with a cycle time of about 8 hours. It was crucial to keep the agitator running, in order to control the exothermic reaction, so a high (top) priority alarm was in place to warn operators if the agitator stopped.

However, at the end of every batch, the agitator was turned off by the process control computer, which activated the same high priority alarm, even though it was not required under these circumstances. With this alarm sounding about every 40 minutes, the operators soon began to ignore it. Despite reporting the matter to management, the issue was not resolved. Operators got so weary of hearing the regular sound of the very loud alarm with no volume control, that they wrapped it in several layers of bubble-wrap to silence it.

Therefore, the likelihood of a rapid and appropriate response to a "stirrer trip" alarm during the reaction phase was drastically diminished.

**Example Incident 3.15 – Batch Reaction Alarms Ignored** *(cont.)*

Lessons learned in relation to abnormal situation management:

- Process Monitoring and Control:

    o Alarm systems and associated procedures must align. Alarms should be meaningful and relevant to the operator in all modes where they receive them.

    o The suppression or isolation of nuisance alarms must be carried out under a management procedure and "bad actors" should be dealt with as detailed in Section 5.3.2.

- Procedures and Design: Operating teams should be involved in designing such systems and associated procedures.

### Example Incident 3.16 – Oklahoma Well Blowout 2018

This case was described in Example Incident 3.4, due to its relevance to abnormal situations associated with management of change and training of personnel on new procedures(2).

This incident concerning the "tripping" of a well, when normal drilling operations had ceased, and the drillers were preparing to remove the drill pipe also has other implications for management of abnormal situations.

An additional key factor was that there was no separate alarm mode for the tripping operation and that many of the alarms had been set up for the drilling mode. The operators had turned off the alarm system that had been activating excessive nuisance alarms, many of which were considered not relevant for this particular mode of operation. A recommendation to address this included:

> *There is also a need for alarm system providers to design the user interface to allow for easy navigation between the state-based alarm operations. Switching between operating modes on the alarm screen should be an easy action for the driller. For example, developing icons on the user-interface screen that the driller can select for the drilling, tripping, circulating, or surface operations—where alarms are already pre-configured—could provide a quick and easy method to switch between the alarm states. (CSB 2019).*

Lessons learned in relation to abnormal situation management:

These are similar to the Example Incident 3.4 lessons, i.e.:

- Process Monitoring and Controls: Alarm systems and associated procedures must align. Alarms should be meaningful and relevant to the operator in all modes where they receive them.

- Procedures and Design: Operating teams should be involved in designing such systems and procedures.

### 3.4.2.6 Type of Operation/Steady-State vs. Transient

Chemical processes may be operated in continuous, semi-continuous, or batch modes of operation. Batch operations, in particular, have a history of incidents precisely because: (a) they are frequently operated across a range of operating conditions within a single batch, and/or (b) reactions frequently have the potential to run away, as cited in Example Incident 3.17.

---

**Example Incident 3.17 – Polystyrene Reactor**

A polystyrene production facility had a history of runaway events that resulted in emergency dumping of reactor contents, and odorous fumes of unreacted styrene dispersing through the surrounding residential area. The fumes were not life threatening, but they were objectionable enough and present often enough that the local authorities demanded a stop to the events.

Given that the normal procedure for handling a runaway was to dump/vent the reactor contents *intentionally* to the atmosphere, a more comprehensive hazard management process was required. First, a HAZOP was conducted to identify all the (many) possible causes of a runaway, then a fault tree was used to quantify the relative importance of each. The site team developed a proposed fix to the problem (installing a pre-release vent pot to capture and cool the vast bulk of the dump), but the fault tree indicated that in 80% of the routes to a runaway, the vent pot would be undersized. Ultimately, the solution was to rely on procedures and systems to *prevent* the runaway rather than *manage* it.

Lessons learned in relation to abnormal situation management:

- Knowledge and Skills: The team recognized the situation, which in this case was routine and not abnormal. They then developed a proposed fix and conducted an analysis of their proposal. This was excellent teamwork.

- Procedures: The team created procedures to prevent the runaway, which is normally a better option than to have to manage the consequences.

---

### 3.4.2.7   Temporary Unplanned Holds or Pauses

Temporary unplanned holds or pauses are not typical of continuous operating plants. When an unplanned temporary hold or pause occurs in a process, whether continuous or batch, this situation is often not addressed by normal operating procedures. This may occur during startup, shutdown, or during normal operations, and it may be related to weather or ancillary equipment failure. Such holds or pauses may occur due to any number of causes. Plant operators need to recognize when such a condition is an abnormal situation that must be managed accordingly. Brainstorming of causes that could create a hold or pause is recommended. Brainstorming exercises may need to include process control and software personnel, as many holds and pauses are software related.

### 3.4.3   Types of Material Being Processed

### 3.4.3.1   Issues Associated with Flammable Operations

Compared to other abnormal situation types, flammable operations pose a greater risk of escalation – a fire leading to a Boiling Liquid Expanding Vapor Explosion (BLEVE), an explosion causing damage to a nearby toxic tank, etc. In these cases, written procedures may need to incorporate an element of emergency response, in order to mitigate the consequences of the abnormal situation.

In the case of a fire that could escalate to a BLEVE, the conventional fire response itself will probably not be in the scope of a unit-based emergency response procedure. Nonetheless, the procedure must include rule sets for evacuating the area for normal unit staff and first responders (and to what distance), as well as provide critical contact information for other affected staff/public. Thus, flammable event response procedures must address not only the immediate unit response (notify first responders, isolate the event) but also any potential escalation outcomes.

### 3.4.3.2 Issues Associated with Various Chemical Phases

Several major industry events have occurred because of personnel not understanding or addressing that an abnormal situation is developing at the interface between different phases of a chemical mixture. Most often this is simply a case of a faulty instrument (e.g., level transmitter failure). However, in other cases, the instrument may be performing correctly but the resulting output is not reflective of the actual situation due to transient conditions. Examples of each are provided in Example Incident 3.18 and Example Incident 3.19.

---

**Example Incident 3.18 - Unreliable Interface Detector**

In a hydrofluoric acid (HF) alkylation unit, the interface detector between the HF and hydrocarbon phases was unreliable, on occasion reading zero level of the hydrocarbon phase when the phase was, in fact, present. One evening this loss of level on the instrument was observed as usual but dismissed by the control panel operator because of the prior history of faulty readings.

After an hour of this operation, the supervisor noted several other atypical issues in the unit – reduction in feed consumption, increase in waste product, strange separator readings, etc. Since these were also symptoms of a zero-level condition, he concluded that the loss of level might be real, and that the unit was actually in an 'acid runaway' condition. The fix for such a situation was to stop the feed to the unit and regroup. However, since neither the supervisor nor the area supervisor had authority to shut the flow, they deferred to the overall plant night shift supervisor, whose decision was to wait for four hours and, if the level was still reading zero, then initiate a shutdown. About six hours after the initial zero level indication, the unit was finally shut down, but because the shutdown had been delayed the restart took three days to accomplish vs. about one hour if it had been initiated immediately. Because storage for the intermediate feedstock had not yet been commissioned, about 25,000 barrels of LPG had to be flared.

---

**Example Incident 3.18 - Unreliable Interface Detector (*cont.*)**

The site had a procedure for a zero-level case, but implementing it was delayed because of uncertainty in the instrument's reading, and a lack of full knowledge on the part of the control panel operator about other clues that could be used to verify the actual state of the level. This demonstrates the need for not just having a procedure but assigning the *authority* for implementing the procedure to the right people and *training* the control panel operator on backup detection of performance indicators.

Lessons learned in relation to abnormal situation management:

- Understanding abnormal situations: The situation was generally understood, but because of unreliable level devices, the response to the situation became an uncertainty.

- Organizational Roles and Responsibilities: This organization had three layers of decision-making involved, which contributed to lagging response. Assigning authority to the front-line personnel can improve response time. "Stop Work Authority" needs to be defined and understood.

- Knowledge and Skills: In this example incident, an increased knowledge and training on other process parameters could reduce the time to diagnose the situation.

---

### Example Incident 3.19 – Phase Concentration

Several industry events have resulted from concentrating a minor component in one area of a distillation tower. In one case, the component was normally innocuous, but in concentrated form, it was explosive. When the tower received a shock (hammer), the component detonated, destroying the tower. In another case, a minor but pyrophoric by-product was concentrated onto one tray of a tower over a period. During the next shutdown, air was unintentionally introduced to the tower, resulting in the tower melting at the tray where the pyrophoric by-product was concentrated.

In these cases, the hazard should have been realized, and a procedure for preventing the undesired condition should have been developed.

Lessons learned in relation to abnormal situation management:

- Understanding abnormal situations: Since there has been a history of similar industry events, learning from previous incidents can be very beneficial.

- Procedures: Written troubleshooting procedures and a procedure for preventing the undesired condition is recommended.

---

### 3.4.3.3 Issues Associated with Different Physical Properties of Chemicals Being Processed

Instrumentation, particularly flow and level instruments, are calibrated based on the physical properties of the substances that are expected to be in the process at that point. Abnormal situations can lead to a different chemical mixture being present that can result in a significant error in the readings from such instruments. This can result in the incorrect diagnosis of situations and has led to several incidents in the past. The BP Example Incident 3.3 (CSB 2007) is one such example, and another example is provided in Example Incident 3.20.

### Example Incident 3.20 – Physical Property Differences

A hydrodesulfurizing unit included an amine scrubber to remove hydrogen sulfide from a hydrocarbon vapor stream. The amine scrubber circulation tank was fitted with a differential pressure level gauge that was calibrated for amine solution that had a specific gravity of about 1.03.

Because of an upstream problem with hydrocarbon separation, liquid hydrocarbons entered the scrubber and the circulation tank. The specific gravity of the hydrocarbons was about 0.6, leading to a significant under-reading of the level in the circulation tank. As a result, the level control system allowed the tank to fill hydraulically to the top, although it was still indicating a normal level. This created additional pressure in the tank that released through safety relief valves. The operators believed that the problem was with the relief valves that were not correctly seating, so they isolated one of them, attempting to reseat the valve. This led to overpressure and major damage to the tank.

Lessons learned in relation to abnormal situation management:

- Understanding abnormal situations: This event occurred because the situation was misunderstood, resulting in the unsafe isolation of the relief valve. The operators had a poor mental model (understanding) of the process, which led them to diagnose the situation incorrectly and the take an incorrect action.
- Knowledge and Skills: Increased knowledge of upstream process upsets that may affect a process is recommended.
- Communications: Increased communications between process units can be helpful in preventing or diagnosing abnormal plant upsets.
- Procedures: A robust management procedure should be in place to ensure relief valves are not isolated without conducting a risk assessment and approval process.
- Process Monitoring and Control: Where processes vessels can contain liquids of varying density, consider a technology that compensates for density variation, or is independent of it.

### 3.4.3.4    *Issues Associated with Chemical Reactivity*

Any number of issues can occur because of chemical reactivity issues, as described in Example Incidents 2.2, 3.1, 3.2, 3.13, 3.15 and Case Study 7.3. Note that the lessons-learned may not be solely about having procedures in place. A more holistic view may be required since incidents associated with chemical reactivity can include:

- Process design
- Process monitoring and control
- Knowledge and understanding of thermal stability
- Recognition of abnormal situations
- Management of change
- Organizational roles and work processes
- Maintenance of critical equipment

# 4 EDUCATION FOR MANAGING ABNORMAL SITUATIONS

The purpose of this chapter is to provide guidance on training for the management of abnormal situations. It is not possible to provide specific training for abnormal situations that will be appropriate for all operating facilities. The content in this chapter is directed not only at site personnel, but also at those who may be remote from day-to-day operations and may have an influence on the occurrence, development, and control of abnormal situations.

## 4.1 EDUCATING THE TRAINER

The content includes advice, tools, and techniques for the provision of such training, both in the classroom, in local workgroups (such as "tool-box talks"), via E-learning, and using DCS simulators or "digital twins". As discussed in Chapter 1, simple computer-based training material is offered in conjunction with this book, consisting of five separate scenarios of abnormal situations that develop on part of an operating process involving distillation, pumps, tanks, heat exchangers and various instrumentation/controls. These training modules can be used by supervisors, plant engineers, and trainers to train operating teams in the diagnosis of an abnormal situation. The material presents background information, develops the scenarios, and provides prompts for trainers and trainees to evaluate what is happening and why. This should enable discussions on abnormal situations and their diagnosis, actions to be taken, learning, and relevance to their operation. This type of material could be developed and extended by management/training personnel to include other situations such as startups, shutdowns and other non-steady state operations specific to their operations. Details on how to access this material is in Appendix A.

The example incidents (such as Example Incident 4.1) also provide useful examples to the various groups that can influence the successful management of abnormal situations.

---

### Example Incident 4.1 – BP Texas City, 2005 (2)

The CSB Investigation Report into the 2005 BP Texas City incident (CSB 2005), (See also Example Incident 3.3, Ch. 3) stated that one of the contributing factors associated with the 2005 BP Texas City incident was that operator training was ineffective.

Lessons learned in relation to abnormal situation management on the topic of training were:

- No effective training materials were created for abnormal situation management.

  o Simulators unavailable.

  o Hazards of unit startup, material-balance calculations, managing tower over-fill scenarios not adequately covered by the training.

- Lack of training on operations technology (distillation principles, level measurement).

- No effective verification methods of operator competency.

- 1998 – 2004 training staff cut from 28 to 8, move to computer-based training, and halving of training budget. CSB found these changes were driven by cost savings.

---

## 4.2    PRIMARY TARGET POPULATIONS FOR TRAINING

The causes of abnormal situations, and the effectiveness of the controls and barriers that prevent them from becoming major events, typically originate in one or more of several key areas, often remote from the facility, including:

- Conceptual design (e.g., process chemistry, material properties, inherent safety)

- Engineering design (e.g., identification/ management of risks through HIRA; plant layout; design of HMIs, inclusion of RAGAGEP, process control strategy)

- Operating and maintenance procedures (e.g., clearly identifying consequences of incorrect actions; steps required to correct or avoid process deviations; provision of checklists or in-hand procedures for tackling infrequent operations; procedures covering all phases of operation and maintenance from startup to decommissioning)

- Abilities and competencies of operating and maintenance staff

- Organizational structure, supervision, and management (including leadership, culture, and operating discipline)

- Plant and equipment reliability

- Effectiveness of active and passive prevention/mitigation systems and emergency response plan/personnel readiness.

Based upon the list of key areas related to abnormal situations, the primary target populations for this training are front-line operators, operations managers, plant engineers, training personnel, process control engineers, and process safety engineers. Asset integrity engineers should also be aware of the concept of abnormal situations when reviewing plant and equipment reliability and while developing integrity operating windows (IOW) programs. Often, the origin, development, and resolution of abnormal situations involves multiple parties who may not be present on the day of the abnormal event. Therefore, management should consider focused training for design engineers, technical experts, and others who might benefit from learning about abnormal situation management. It is also recommended that multiple workgroups are involved in training and simulations as further outlined in Section 4.3.

An excellent reference for effective training is *Guidelines for Defining Process Safety Competency Requirements* (CCPS 2015). Chapter 2 of the book covers the generic roles of an operational production unit and the normal expected training competencies for each role. The next section further discusses roles and training with respect to managing abnormal situations.

## 4.2.1   Front-line Operators

Process operation during abnormal situations can create a high-pressure environment for the operators. Efficient management and the correct handling of the situation are key in preventing its escalation into a more serious incident. Sudden, potentially dangerous situations can affect human performance (the "startle" factor), leading to a "fight or flight" response, that can lead to inappropriate action being taken. Non-technical skills training on how to perform effectively in stressful situations could be highly beneficial. The aviation industry refers to Crew Resource Management training, which can improve team performance under such situations. This is discussed further in Section 5. Control room operating personnel (panel or board operators) must have the competencies and skills to be able to:

- Identify the initiation of an abnormal situation.

- Apply their mental model of the process to understand the situation and formulate a plan to identify the appropriate action(s) that may be required.

- Troubleshoot the cause of the abnormal situation calmly and thoroughly, either directly or in association with another member of staff such as an outside operator, plant engineer, supervisor, manager, or perhaps a subject matter expert.

- Understand the process safety issues associated with shutting down or isolating equipment (e.g., trip overrides), including requirements for temporary management of change (MOC) and pre-startup safety reviews (PSSRs).

- Restore the situation to normal or make further decisions such as whether to continue operating, reduce production rates, or to shut down the process.

Outside operators (field operators) require similar abilities albeit in a different environment. They will likely be outside, potentially in poor weather, at a noisy and possibly poorly lit area with many distractions and the only communication available might be a hand-held device such as a walkie-talkie, transceiver, or radio. Under such conditions, they will often be called upon by a control panel operator to help diagnose a situation such as a stuck valve, faulty level indicator, cavitating pump, local pressure gauge reading, or pump trip. Outside operators should be vigilant and alert to changes in the work environment during their routine field activities that could indicate an abnormal situation. Operators should note such changes in order to further investigate and understand their significance, and they should report them as necessary to their supervisor. Examples of changes that may require further investigation include:

- The behavior of the flame in a furnace or firebox that could be a sign of low or high fuel gas pressure.

- Changing color and smoke from flare stack.

- Hearing abnormal noise or observing excessive vibration in rotating equipment

- Unusual smell

- Pooling of liquid or visible gas

Note: Training for outside operators is discussed further in Chapter 5 Section 5.6.

Front line operators are often called to be part of a HAZOP/LOPA (Layer of Protection Analysis) team and their active participation in the structured discussions on the practical aspects of handling deviations from normal operating parameters are crucial, see 4.2.4.

Ongoing developments in connected devices have been helpful, such as live-streaming from the field to the control room or to a subject matter expert (SME). Other connected devices that measure parameters such as sound and vibration can help to diagnose problems remotely.

## 4.2.2 Operations Management

Operations supervisors and managers require many of the competencies and skills of the operating personnel, and they require good leadership skills. Through this they build a culture of trust, mutual respect, and cooperation within and between the operating teams—another key factor in the successful management of abnormal situations. Since operations managers could rotate to new positions on a regular basis, they have limited time to build significant knowledge of the process and gain historical experience of upsets and precursors to abnormal situations. Operations managers usually recognize this and readily acknowledge that more experienced front-line operators and supervisors are entrusted to make initial responses to abnormal situations.

In addition to training in management and leadership, operations leaders require further training for management of abnormal situations including the emergency response roles for themselves and their team. This includes effective delegation of critical decisions to the operating teams such as "Stop Work Authority" (i.e., there will be no criticism or reprisal if operating personnel decide to shut down the process due to perceived safety issues that are ultimately discovered to be unfounded). Managers also need to know and understand the decisions they are required to make and to ensure there are accurate and reliable lines of communication between both them and senior management as well as the operators and supervisors. Managers should also make clear the hierarchy of decision-making and associated procedures for abnormal operations involving the shutdown or isolation of certain plant or equipment. For example, an operator might be authorized to shut down a piece of equipment if it is not safety-critical. However, if it does affect safety critical equipment, managers must ensure that procedures for temporary operation (such as bypass of safety devices) are understood and followed, including who needs to be notified and what other approvals are needed if the appropriate level of supervision/ management cannot be reached.

Operation managers are typically the highest-ranking personnel in the production unit and may delegate many responsibilities to shift supervisors or shift leaders as a first level of the management team. Thus, these supervisors and leaders also require many of the same skills and competencies expected of the higher-level management personnel. Management should ensure that specific roles and responsibilities are delegated effectively for circumstances when key operations management personnel are absent.

### 4.2.3    Plant Engineers/Technicians

Plant Engineers, including mechanical, instrument/electrical and process control/automation engineers and technicians also have an important role in management of abnormal situations that could include:

- Ensuring that critical equipment is designed and performing correctly.

- Identifying hazards resulting from planned changes.

- Prioritizing repairs of critical equipment.

- Assisting operating personnel with diagnosing and troubleshooting instrument and equipment faults.

- Providing expert advice during an abnormal situation on equipment functionality and failure modes.

- Rectifying faulty equipment or providing temporary workarounds (In association with temporary Management of Change (MOC) procedures).

### 4.2.4 Process Safety Engineers

Process Safety Engineers at a manufacturing site generally assume a support role in identifying possible sources of risk in process areas that may require additional assistance in managing abnormal situations. After an event, the Process Safety Engineers will seek to understand the causes and contributing factors that led to the abnormal situation, ensure the learning is shared across the manufacturing site as well as the company, and assist in the identification of mitigation measures and specific training to prevent a recurrence. They will also ensure that any action items are tracked and monitored, the responsible manager follows their action(s) to completion, and where required, the learning is embedded into systems and procedures. In a more remote role, such as in engineering design or other technical role, their involvement may be through HIRA, including HAZOP and LOPA. As mentioned in 4.2.1, such studies should involve front-line operating staff, who can provide input based on their practical experience.

HAZOP/LOPA studies typically consider causes and consequences of a single deviation of an operating parameter such as "high level" or "low flow" based on a guide word approach. While an abnormal situation may initially be a consequence of a single deviation, typically this escalates into a series of deviations that can sometimes hide the initial issue. Consideration of how a single deviation if unchecked, could evolve into a more complex situation (that could be considered "abnormal") should form part of the scope of a HAZOP. Recommendations following the study could include changes to process design, procedures, and controls/ instrumentation, as well as developing a strategy to handle such a series of deviations including "first out alarm reporting" and "alarm suppression" that are discussed further under Section 4.2.5.

### 4.2.5 Design Engineers

A major factor in the ability of operating and maintenance personnel to respond correctly to abnormal situations is the design of the process and the Human Machine Interface (HMI). The control room operator's view of the process operation is significantly affected by the design of the HMI. Effective operator HMI design practices that establish a high level of operations team situational awareness is one way for operating companies to enable a proactive operating posture. (Bullemer & Reising, Effective Console Operator HMI Design (2nd Edition), 2015). This is also the case for the outside (field) operators who interface/interact with the valves, gauges, and other pieces of equipment. Design engineers normally play an important role in recommending the design and layout of the console displays, alarm systems, field equipment and almost all other features of the HMI. Therefore, design engineers should be made aware of the management of abnormal situations, which should be included in their training sessions. This is especially important when alarm structure and display are being considered. Some published RAGAGEP standards such as those by ISA, EEMUA, and IEC (ANSI/ISA 2009; EEMUA 1999/Revised 2013; IEC 2014) should be applied when appropriate.

In addition to applying these RAGAGEP standards, several techniques that can be employed in the design of the HMI include:

- Alarm configuration such as "First out alarm reporting" is key to understanding the initial alarm that directly indicates the presence of a developing abnormal situation. It is, by definition, what alarm cascade management is intended to solve.

- Alarm management strategy such as "alarm flood suppression" is the ability to support alarm cascade management by limiting initial alarms to those that are meaningful and actionable. For example, if a high-high level trip activates on a compressor inlet knockout drum, shutting down the compressor, the operator only requires an alarm indicating the high-high level and the trip. The operator does not need subsequent alarms for high suction pressure, low discharge pressure, low lube oil pressure, and all of the other alarms that would accompany a compressor shutdown, which can make it harder for the operator to determine what happened and what to do next. This strategy is typically applied to complex process equipment systems such as refrigeration systems, burner management systems, and self-contained packaged units.

The design and layout of plant and control equipment in the field is also highly relevant to the effective management of abnormal situations. The logical positioning of controls and valves relative to the items they affect will help to ensure that the correct device is operated in an emergency. The clear and correct labelling of equipment, locally visible valve position indicators, and marking of the instrument range on field gauges will help the field operator in their troubleshooting and mitigating actions. Additionally, the instrument design engineers and process engineers must also ensure that the ranges of the instruments cover all potential operating modes and scenarios.

Further tools and techniques on HMI design are discussed in Chapter 5, Section 5.7.1.

If a process can be designed to be inherently safe, then this is always the preferred option, as discussed in Example Incident 4.2.

---

**Example Incident 4.2 – Chernobyl Disaster, April 1986**

One feature of the design of the RBMK nuclear reactor at Chernobyl is that it was primarily graphite-moderated and cooled by water. However, water is also a neutron moderator and therefore when allowed to boil within the reactor, the void created by the steam results in a reduction in moderation. This creates a thermal feedback loop where more power creates more boiling and less moderation, which in turn leads to more power. The condition is called "positive void coefficient" and the reactor design had the highest positive void coefficient of any commercial nuclear reactor.

Once the water in the core started to boil beyond a certain rate, the operators could do very little to control the reaction.

Many other factors were associated with this incident, although a major design feature of the nuclear reactors in other industrialized countries is that they are designed with negative void coefficients, which makes them inherently safe from this type of runaway situation.

**Example Incident 4.2 - Chernobyl Disaster, April 1986** *(cont.)*

In addition to the design issues, lessons learned in relation to abnormal situation management are:

- <u>Abnormal Situational Awareness</u>: The cooling system of this particular reactor system was not designed to mitigate an inherent risk of runaway core heating. The conditions that could lead to this heating situation should have been understood by operating personnel.

- <u>Procedures</u>: Written procedures for safely managing and preventing abnormal situations involving the cooling process should be considered safety critical procedures.

- <u>Knowledge and skills</u>: Front-line personnel should be fully trained on all procedures.

Simple design features can, particularly under stressful conditions, lead to an incorrect response by an operator. Such incidents may initially appear to be caused by "human error", although a thorough root cause investigation often reveals underlying design or management system issues. Current Example Incident 4.3 (Syed 2015) shows this type of error.

---

**Example Incident 4.3 – Boeing B-17 Bomber, 1940s**

The Flying Fortress suffered from numerous runway crashes that no amount of training ever seemed to fix. These were considered to result from "pilot error," but it was not until after the end of WW2 that the solution was found. Psychologist Paul Fitts and his colleague Alfonse Chapanis established through interviews with the pilots that the controls for the flaps and the landing gear looked and felt exactly the same. Under relaxed flying conditions, this was not a problem. Under the pressure of a difficult landing, however, the pilots were pulling the wrong lever. Instead of lowering the landing gear, they were pulling the wing flaps, slowing their descent, and driving their planes into the ground with the landing gear still up.

Chapanis came up with a simple solution, using different shapes attached to each lever, so that they could be easily distinguished, even in the dark. The levers now had an intuitive meaning, easily identified under the pressure of a difficult landing.

In addition to the design issues, the lesson learned in relation to abnormal situation management was:

- Monitoring: The cockpit of an airplane could be compared to a control panel. In this case, the design of the control led to human performance issues under stress. The abnormal situation aspect of monitoring focuses on effective design, deployment, and maintenance of hardware and software platforms that support process monitoring, control, and support for effective operations, thus lessening the chance of human performance issues during stress.

---

Prior to the introduction of computerized control systems, analogue control rooms provided vertical panels, typically arranged by process unit, that the control room operators could walk around. Mimic panels were used to display more complex parts of the process pictographically and instrument alarm panels were typically shown above, with colors corresponding to the alarm priority. A good example of an old style of control room is one operated by NASA in the late 1940s, shown in Figure 4.1 (NASA).

**Figure 4.1 NASA Control Room – Engine Research Building**

A well-designed analogue control room would provide operators with an overall "feel" for the process that was, to some extent, lost when digital control systems were first introduced.

In the late 1970s and early 1980s when DCS systems were first used in the process industry, display screens were large (deep cathode-ray tubes) and expensive, and it was impractical to replicate the layout of the previous analogue control rooms. Color printers, to obtain hard copies of trend data were also expensive and some operators even used instant-print cameras to take screen pictures so that trends could be examined. The DCS systems provided major benefits, but some early systems were not well designed for the operators and failed to provide ready access to alarm screens and system overviews. It was often necessary to page through several windows before the required information could be obtained, by which time the original window had been closed.

Another feature of the DCS system with both positive and negative aspects was that users were able to add an alarm to the system at zero cost. This led to a large increase in the number of alarms and several incidents where alarm overload (flood) was a significant contributing factor.

Recognizing alarm overload led to the introduction of the principles of alarm rationalization and alarm management, which considered human factors/ergonomics on the part of the operators and supervisors.

Various standards and guidelines on alarm management include:

- ANSI/ISA-18.2, "Management of Alarm Systems for the Process Industries" 2009, revised 2016 (ANSI/ISA 2009/Revised 2016)

- NA 102, "Alarm Management", Issued by the User Association of Process Control Technology in Chemical and Pharmaceutical Industries (NAMUR), 2003 (NA102 2003)

- EEMUA 191 "Alarm Systems- A Guide to Design, Management and Procurement", 1999, revised 2013 (EEMUA 1999/Revised 2013)

- API 1167, "Pipeline SCADA Alarm Management", 2010, revised 2016 (API 2010, Revised 2016)

- IEC 62682, "Management of alarm systems for the process industries", 2014 (IEC 2014)

Further details on alarm rationalization and console design are provided in papers by Emerson, Rockwell, and ASM© Consortium (Emerson 2019, Rockwell 2017, Bullemer & Reising 2015).

Process automation engineers and process engineers providing input on alarm configurations must be knowledgeable concerning these standards and guidelines to ensure the operating teams:

- Are not overloaded with alarms.

- Can easily identify alarm priority (e.g., high priority safety alarm or process alert).

- Are presented only with alarms that are meaningful and actionable based on the process or equipment state or operating mode.

- Can select an overview of each section of the process, and the overall unit, as required.

- Are able to identify the initiating alarm for a particular event and the required action to be taken.

Human factors are considered a major part of alarm management and design of the HMI. Example Incident 4.4 comprises an event where the operators (pilots) were unable to evaluate what has happening under conditions of high stress (French BEA Final Report).

---

**Example Incident 4.4 – Air France AF 447 Crash, June 2009**

On June 1, 2009, Air France flight AF447 (Airbus A330-203) flying from Rio De Janeiro to Paris crashed into the Atlantic Ocean approximately 3 hours and 45 minutes after takeoff. The accident resulted in the fatality of 228 passengers and crew members. The French Bureau of Enquiry and Analysis for Civil Aviation Safety (BEA) investigated the accident and released the final report three years after the fatal crash.

The report identified the blockage of pitot tubes responsible for speed measurement as the first of a series of events that led to the accident. On an aircraft, three sets of pitot tubes are used to determine key flight parameters including speed and altitude. Blockage of the pitot tubes caused inconsistencies in aircraft speed measurement that resulted in disengagement of the autopilot leading the airplane to a stall position. The crew failed to recover from the stall position.

According to the investigation report, "The blockage of Pitot probes by ice crystals in cruise was a phenomenon that was known but misunderstood by the aviation community at the time of the accident. After initial reactions that depend upon basic airmanship, it was expected that it would be rapidly diagnosed by pilots and managed where necessary by precautionary measures on the pitch attitude and the thrust, as indicated in the associated procedure. The crew, progressively becoming de-structured, likely never understood that it was faced with a "simple" loss of three sources of airspeed information."

**Example Incident 4.4 – Air France AF 447 Crash, June 2009** *(cont.)*

The report goes further to identify several contributing factors related to crew recognition and management of the situation, including:

- Incorrect actions taken by the crew upon auto pilot disconnection destabilized the flight path.

- Failure of the crew to initiate procedures upon losing flight speed.

- Failure of the crew to recognize the stall position in a timely manner.

This incident prompted increased reporting from airline operators of similar problems with pitot tubes in heavily icing conditions and led to a prohibition of certain models of probes as a precautionary measure. In addition, the maintenance interval for pitot cleaning was reduced.

Lessons learned in relation to abnormal situation management:

Among the 25 safety recommendations issued by BEA, the following were made with regards to crew instruction & training:

- Knowledge: Improve crew knowledge of aircraft systems and changes in their characteristics in degraded or unusual situations

- Skill Development and Training: Improve flight simulators for a realistic simulation of abnormal situations.

### 4.2.6    Environmental Health, Safety and Security (EHSS) Personnel

EHSS personnel are more likely to act as responders to the consequences of abnormal situations that have developed into a loss of primary containment. Events that are more significant typically require a response by emergency teams with outside industry assistance and external responding agencies.

### 4.2.7    Technical Experts

Technical experts, also called "subject matter experts" (SMEs) are often located remote from the facility but could be called upon to help to diagnose or respond to an abnormal situation (as discussed in Example Incident 4.5). While it is not possible to consider the whole range of technical expertise that may be involved in such a situation, these individuals could include:

- Instrument engineer, to explain how a level gauge may not respond to a change in level.

- Rotating equipment specialist, to help diagnose why a compressor starts to vibrate under abnormal conditions or the rotating equipment vibration profile has changed over time.

- Quality control engineers/specialists to help troubleshoot contamination and composition of raw and process chemicals.

- Materials engineer, to help understand the significance of a minor crack that appears in a flange on a pressure vessel.

- Piping engineer to evaluate the risk associated with the movement of a pipe support.

- Technology Licensors to provide specific expertise related to the process.

**Example Incident 4.5 – Chlorine Pipeline Support - 1992**

A facility that both manufactured and consumed chlorine was subjected to 3 days of intense rain. The following weekend, an operator observed that a concrete post beneath a liquid chlorine pipeline had shifted due to a minor landslip such that it no longer provided support for the pipe, as intended. The issue was raised and escalated until it reached the works manager who immediately issued instructions to shut-down several plants and depressurize the pipe until expert advice could be sought.

Piping and Civil Engineers were called to evaluate the stresses on the pipeline and the options to provide the required support. They concluded that the pipeline had not been subjected to a significant increase in stresses and that an additional support could be designed and erected expeditiously. Two days later the work was complete, the pipeline was pressure tested and the systems were put back into service.

The involvement of SMEs was critical in assessing and rectifying this abnormal situation, which no one on the site had dealt with before.

An additional outcome of this incident was a demonstrated commitment from top management to matters of safety that helped reinforce the safety culture already being built on the site.

Lessons learned in relation to abnormal situation management:

- <u>Understanding Abnormal Situation</u>:  In this situation, the field operator observed an abnormal piping support and communicated the finding to plant personnel, escalating to unit leadership. This was excellent and exactly what abnormal situation management strives to achieve. A potential serious pipeline failure was avoided.

### 4.2.8   Other Parties

Other individuals who could potentially be involved in resolving abnormal situations are as follows:

- Laboratory technicians, who are able to identify or characterize issues with intermediate products as part of routine or special troubleshooting analysis.

- Incident command center, where key decisions are typically made including how to deal with losses of containment, involvement of external resources, mutual aid, communication with higher management and local external parties.

- Corporate Headquarters, where major strategic decisions are made on handling significant events and communication with the media, shareholders, and other interested parties.

## 4.3   GUIDANCE FOR ORGANIZING AND STRUCTURING TRAINING

### 4.3.1   Organization of Training

Training workers and assuring their reliable performance of critical tasks is one of the nine elements in the RBPS pillar of managing risk. Establishment of a training management system is the initial step, key elements of which will normally include objectives, measurements, training materials, and effective trainers. This approach is generally accepted as the fundamental basis for most programs, although the objectives may vary greatly across industries and occupations. For example, the objectives of a training program for astronauts would be different from training for front-line plant operators. However, both need to understand the functioning details of the equipment and control systems; and both need to know how to recognize and respond to abnormal situations.

Training is often conducted within specific workgroups, such as the operating team, the maintenance team, or the engineering department. For abnormal situation training, however, it may be more appropriate to adopt a holistic or "systems-thinking" approach when conducting at least some of the training, and to involve various workgroups and experts.

This type of arrangement works well for HAZOP studies and has the added benefit of providing feedback to designers and engineers who are not on the front-line and may not have a full appreciation of potential issues with the equipment they are providing. A common criticism offered by operating personnel is that the designers do not have to run the process and therefore sometimes fail to have an appreciation of some of the problems that certain design features can create.

Traditional training and learning processes are normally targeted at teaching personnel to operate the process per standard operating procedures (SOPs), step-by-step tasks or checklists, and using the computerized process control system. The training tools that are used are typically very structured to ensure consistency in how the training information is communicated and how understanding is verified. Although these types of tools have been demonstrated to work well, they may not be the optimum tools for educating for abnormal situations. These matters are discussed further in Chapter 5.

No matter the training objective, logistics of the training session including time, setting, frequency, and number of personnel on shift, as well as type of training, (e.g., Classroom / Computer/ Desktop / One-on-One) should be considered in advance. Management of the training should include a system for enforcement of the training and metrics to measure the effectiveness of the program, as discussed in 6.2. The next section describes some of the training topics and structure.

## 4.3.2    Structure of Training Topics

### 4.3.2.1    *Basic Process Operations*

For front line control room operators and field operators, an introductory understanding of the process chemistry, potential chemical interactions, relationships between temperature, pressure, flow, and level as they relate to process operations are essential to building an operator's knowledge and the ability to recognize and respond to abnormal situations. Training in these basic relationships can be accomplished using basic process simulation software that typically runs on a desktop personal computer. Additionally, the training may be accomplished in informal training sessions with senior plant personnel or in formal training with chemical process instructors.

In addition to basic training, operations and maintenance trainers can use their own experience to develop scenarios and relate situations that can be used for operator and maintenance technician training. These are often the best-remembered training examples since they tend to be real life related and based on actual experience. The training should include troubleshooting to provide operators with the skills, knowledge, and resources to diagnose faults with equipment such as instrumentation, machinery, and processing equipment. Informal, local discussions to test knowledge and understanding helps keep the team engaged and adds to the quality of the training. The training could be augmented with videos, animations, training simulators or even virtual reality, to capture the real essence of the actions taken and timely diagnosis made, see Section 5.6. Further advice on non-technical skills training including decision-making under pressure and how best to replicate these in simulations is available in the paper: *Training Decision Makers – Tactical Decision Games* (Crichton et al 2000).

As process plants become more reliable and inherently safer with fewer shutdowns, loss of containment events, fires, and explosions, there are fewer opportunities for plant personnel to experience and learn from such events. Thus, the skills and real-life experience of seasoned trainers has become ever more important.

### 4.3.2.2   DCS Controls, Display Screen and Alarm Management

The importance of correctly assessing and responding to high priority alarms is critical to the process of managing abnormal situations. The design of the operator control consoles was discussed previously especially with respect to Human Machine Interface (HMI). Management of the alarms is often as important as monitoring process conditions, adjusting process control settings, or starting/stopping equipment. Alarms are typically the first indicator of a process that is experiencing an issue. Learning and understanding the alarm system, the structure of high to low alarms including those alarms considered safety critical, and the recommended responses to all alarms is expected for an organization that has an active process in place for the management of abnormal situations. ISA-18.2 (ANSI/ISA 2009/Revised 2016) provides guidance on categorizing alarms and for the training requirements for critical alarms, referred to as Highly Managed Alarms (HMAs). Required training for these HMAs include:

- Initial and refresher training that is appropriate to the HMA
- Maintenance training that covers the specific maintenance requirements for the selected HMA

### 4.3.2.3 Emergency Procedures

Some abnormal situations in the plant or process could result in the need to initiate an emergency response. Training on historical emergency situations, prepared and proven emergency procedures, and establishing a line of communication is therefore recommended.

With this in mind, responding to emergencies is typically less effective if the response requires locating, reading, and executing an emergency procedure. Best practice involves operating teams conducting drills on emergency procedures during off-shift hours. The drills are typically scheduled monthly, or more often, so that all emergency procedures are drilled at least once per year. A typical drill could consist of an experienced operator leading the shift team through a brainstorming session that uses team input to recreate the emergency procedure without actually referring to it. A scribe records the shift team interaction and creates a step-by-step emergency procedure that, at the end of the exercise, is then compared to the actual procedure. The team then discusses any steps that were missed or added, and the updated procedure is provided to the operations trainer for further review and critique.

Each shift team completes a similar exercise for each emergency procedure. The scribed procedures from each shift are collected by the operations trainer and reviewed for new and appropriate or missed steps. At the end of the exercise for each emergency procedure, the procedure has been:

- Drilled from memory
- Reviewed for missing steps
- Reviewed for correct order
- Reviewed for adequate time to respond
- Revised based on all operator teams' inputs
- Applied MOC process to ensure the revised procedure is formally updated and recorded.

### 4.3.3   Skills and Competencies of Trainers

Training may be provided by a variety of means, but the objective is always the same—to help plant operators to be successful in meeting their assigned responsibilities. Training tools and techniques are discussed further in Section 5.6. Success can be measured by the number and severity of near-misses or incidents that occur, the number of emergency situations, and the number of abnormal situations that have occurred. More details of leading and lagging indicators are provided in Chapters 5 and 6.

The quality and success of the training program will most likely depend on the personnel providing the training. These personnel are typically assigned as Trainers across a plant site or within a specific process unit. They may report to a central training organization or a local process unit leader. The success of the Trainer often depends upon that person's formal training, years of training experience, knowledge of the chemical process, interpersonal skills, willingness to share but also listen, and measurement of training progress against objectives. Depending on the size of the facility, the organization's personnel configuration, and the available staffing, trainers may be assigned additional tasks. Therefore, it is recommended to have the trainer's role, responsibilities, and objectives explicitly documented. Any changes to the trainer's documented scope could affect the overall success of the training program and should be considered and managed via the MOC process.

With respect to management of abnormal situations, the training curriculum should include, at a minimum, the topics that are covered in Section 4.3.2.

## 4.4   SUMMARY

This chapter has provided guidance to organizations in the process industries about organizing and structuring training on management of abnormal situations, including advice on trainer competencies, training programs, and focused training topics.

Chapter 5 will introduce the reader to tools and methods for management of abnormal situations including the range of abnormal situation control options. Section 5.6 provides advice on training and drills.

# 5 TOOLS AND METHODS FOR MANAGING ABNORMAL SITUATIONS

Previous chapters in this book stress the importance and value of recognizing and managing abnormal situations in chemical processes and provide many actual example incidents to fortify those lessons. Several management tools have been mentioned, with focus on preventing or minimizing abnormal situations. This chapter further illustrates and provides guidance on applying these tools for managing abnormal situations when they occur. Process plant operators are familiar with many of these tools and often use them in their normal job responsibilities. However, by increasing their knowledge about the tools and their importance, it is possible for the operators to identify opportunities to improve performance on the job by proper use of those tools and methods, especially during an abnormal or stressful situation.

There are several available tools and methods to help predict and/or identify abnormal situations, and thus prevent them from escalating to serious or major incidents. These tools are usually well recognized in the process industry, but with advancements in computer applications and technologies, new tools and methods are continuing to evolve that are likely to provide further risk reduction in the future. This book will primarily cover the existing and currently available tools but will also discuss some new technologies as appropriate.

This chapter is structured to cover tools and methods that are associated with the eight subject areas that are listed next in Section 5.1. These tools and methods are then discussed in more detail in Sections 5.2 through 5.9 in this chapter.

## 5.1    TOOLS AND METHODS FOR CONTROL OF ABNORMAL SITUATIONS

Much of the available literature on the management of abnormal situations focuses on Human Machine Interface (HMI and procedural issues, and to some extent hazard identification (HAZID)) techniques to identify those scenarios that should be considered in an HMI or procedure analysis. Journal articles by Errington, Bullemer, and Ostrowski describe the importance of each of these topics (Errington et al 2005; Bullemer et al 2010a; Ostrowski & Keim 2010). However, as illustrated by the wide range of real-world example incidents introduced in Chapter 3, abnormal situations are clearly not limited to only HMI and HAZID-related events. Therefore, this chapter takes a more holistic approach, considering a broader number of subject areas, arranged into these eight areas:

- Predictive Hazard Identification

- Process Control Systems

- Policies and Administrative Procedures

- Operating Procedures

- Training and Drills.

- Ergonomics and Other Human Factors

- Learning from previous Abnormal Situation Incidents

- Change Management

Most of these subject areas are similar to the ASM® Consortium research areas described in Chapter 3, Section 3.1.1. Each subject area is discussed separately, along with associated tools/methods and links to some of the example incidents from Chapter 3 noted as applicable.

Table 5.1 provides a summary of this chapter, along with references to applicable example incidents from this book, to enable the reader to find examples of each of the areas easily.

**Table 5.1  Abnormal Situation Subject Areas, Tools and Methods**

| Subject Areas (Section) | Tools and Methods | Example Incidents |
|---|---|---|
| Predictive Hazard Identification (5.2) | • HAZOP<br>• What-If analysis<br>• Checklists<br>• FMEA<br>• LOPA<br>• Modified checklists for ASM®<br>• Bow Ties | 2.1<br><br>3.1   3.10<br><br>4.2   4.3   4.4   4.5<br><br>5.3   5.4<br><br>6.1   6.2 |
| Process Control Systems (5.3) | • Monitoring of process trends<br>• Alarm management<br>• Forward-predicting software algorithms<br>• Pattern recognition and machine learning | 2.3<br>3.5   3.11  3.13<br>3.15  3.16<br>4.1   4.4<br>5.1   5.3   5.4<br>5.5<br>6.1 |
| Policies and Administrative Procedures (5.4) | • Daily communications<br>• Situational dialog<br>• Handover shift activities and analysis<br>• Process metrics | 3.2   3.3   3.5<br><br>3.8   3.11  3.13<br><br>3.14  3.18  3.20<br><br>4.1   4.2<br><br>5.2   5.3<br><br>6.1   6.2 |
| Operating Procedures (5.5) | • Accessible and accurate written procedures<br>• Emergency response<br>• Clear and complete startup and shutdown steps<br>• Procedure life cycle management | 2.1<br><br>3.1   3.3   3.5<br><br>3.6   3.7   3.9<br><br>3.12  3.13  3.14<br><br>3.15  3.16  3.17<br><br>3.19  3.20<br><br>4.1   4.2   4.4<br><br>5.1   5.4<br><br>6.2 |

**Table 5.1 Subject Areas – Tools – Methods** *(cont.)*

| Subject Areas (Section) | Tools and Methods | Example Incidents |
|---|---|---|
| Training and Drills (5.6) | • Continuous learning environment<br>• Process simulation<br>• Tabletop scenarios/drills<br>• Toolbox discussions<br>• Learning from mistakes<br>• Causes of previous incidents<br>• Historical industry incidents<br>• Field observations | 2.2<br>3.2  3.4  3.5<br>3.7  3.10  3.17<br>3.18  3.20<br>4.1  4.2  4.3<br>4.4<br>5.3<br>6.1  6.2 |
| Ergonomics and Other Human Factors (5.7) | • Evaluate control room conditions during off shifts and during periods of stress<br>• Ergonomic physical assessment conducted of control room<br>• Human Machine Interface (HMI)<br>• Human performance consequences<br>• Responsive maintenance of control system hardware and software<br>• Control screen displays and alarms assessment<br>• Control screen management during crisis<br>• Emergency process situation, obstacles, and safety systems availability | 3.11  3.12<br>4.3 |
| Learning from Abnormal Situation Incidents (5.8) | • Incident investigation process that includes performance of barriers (worked as designed or failed to work as designed)<br>• Near-misses investigated and reviewed for abnormal situations.<br>• Reference list of relative historical incidents within the company and across the industry.<br>• Process Safety Incident metrics | 3.9  3.19<br>5.2 |

**Table 5.1 Subject Areas – Tools – Methods** *(cont.)*

| Subject Areas (Section) | Tools and Methods | Example Incidents | | | |
|---|---|---|---|---|---|
| Change Management (5.9) | • Management of Change (MOC) procedure requiring operators' involvement.<br>• MOC requirement for changes to the process control system and logic<br>• MOC approval for bypassing safety critical devices, interlocks, safety system alarms, and the opening manual bypasses or closing of block valves in front of safety devices.<br>• Organizational changes (MOOC)<br>• PSSR process | 2.2<br><br>3.2   3.3   3.4   3.6<br>3.9<br>5.1   5.2   5.5 | | | |

## 5.2   PREDICTIVE HAZARD IDENTIFICATION

Table 5.2 provides an overview of some of the most frequently used tools for hazard identification and their corresponding strengths and weaknesses.

### Table 5.2  Hazard Identification Tools

| Common Tools and Methods | Strengths | Weaknesses |
|---|---|---|
| HAZOP, FMEA | Systematic approaches to predict events based on potential scenarios and identify the existing and/or required safeguards. | Resource-intensive |
| What-If Structured What-If, Checklists | Focuses on known prior failure mechanisms or team experience. Often less time is required, as questions are more relevant to the process being evaluated. Can be easily modified for reviewing abnormal situations. | Strongly dependent on the expertise of the checklist developer and the applicability of the checklist to the types of hazards involved in the process being reviewed |
| Event Trees, Fault Trees, Bow Ties | Provides detailed path from cause to final outcome, each branch of which can potentially be a point that mitigation can address. | Logic and mathematics can be easily misapplied by someone who is not a specialist in the technique. |

### 5.2.1 Hazard Recognition for Abnormal Situations

The purpose of evaluating potential abnormal situations is to identify and document the potential hazards and their consequences for each operating phase. This assumes careful analysis of the following:

- Process technology/design and the associated process hazards.

- Inherently safer design features that if applied could reduce to a safe level or even eliminate the process hazards.

- Process control strategy for maintaining normal operations and responding to process upsets.

These activities are should be evaluated for abnormal situation management, although inherently safer design is better addressed before the details of ASM® are evaluated for implementation.

### 5.2.2 HIRA Approach to Hazard Prediction

A Hazard Identification and Risk Analysis (HIRA) approach is typically followed for analyzing the process. Some example tools used within HIRA are **HAZOP Studies, What-If Analysis, Checklists, Failure Modes and Effects Analysis (FMEA), Fault Trees, Bow Ties and Layer of Protection Analysis (LOPA)**. Three recommended CCPS books for learning more about these tools are *Guidelines for Hazard Evaluation Procedures, Guidelines for Risk Based Process Safety*, and *Layer of Protection Analysis-Simplified Process Risk Assessment* (CCPS 2008b, 2007a, 2001).

These tools have been utilized and proven for years in identifying potential hazards and then reducing or preventing their likelihood by recommending the addition of various safeguards and protection layers. However, these tools are relatively standardized to address situations such as high flow, low level, high pressure, seal failure, reaction runaway, or contamination, and may not always consider abnormal situations resulting from scenarios such as:

- Signal from sensor or pneumatic impulse line is interrupted while continuing to indicate normal operating range.

- Heat tracing fails or insulation missing.

- Valve fails to open or close on demand.

- Pressure relief valve sticks open.

- Controller does not respond quickly enough to a process disturbance.

- Drive motor fails.

- Control system console display fails.

- Loss of communication between field devices and controllers.

- Control I/O (Input/ Output) device malfunctions due to major disturbance such as water entering the cabinet.

- Control panel operator places process control loop in manual mode, overriding automatic controls.

- Operator incorrectly bypasses a safety device or function.

- Instrument technician fails to re-open an isolation valve under the field instrument.

- Control changes are poorly communicated to operating personnel.

- Sudden loss of power  (e.g., due to severe weather, lightning, hurricanes, etc.) forces a switch to emergency control power, or to operating a unit-wide backup power supply.

These situations happen and in many cases are not readily recognized, especially the first time they occur within a process. Therefore, for abnormal situations that may affect the process control system, the HIRA process facilitator should encourage the risk analysis team to brainstorm and document historical abnormal scenarios and to stimulate discussion on any additional scenarios that might occur. The HIRA analysis may conclude that some of these abnormal scenarios could result in an unwanted consequence that is not effectively prevented by existing safeguards. For those scenarios, the analysis team can recommend that additional safeguards and procedures be considered.

This 'supplemental' abnormal situation review could be performed as an integrated part of the HIRA. Alternatively, the abnormal situation review could occur after the recommendations from a traditional HIRA review have been addressed to provide clarity on the final design basis for the review. The review can then use traditional HIRA approaches and/or checklists of topics specific to abnormal situations such as the bulleted list in this section.

For established chemical processes, with an experienced operating team in place, another approach to discussing potential abnormal situations is through tabletop exercises or drills. The facilitator can challenge the operating team members to document how they would respond to various upset situations. This type of exercise can be expanded to stimulate further discussion about situations where other abnormal conditions could be encountered. The result of these tabletop drills should be used to improve operating procedures, training of personnel, and installation of additional safeguard controls and hardware.

In summary, traditional HIRA reviews typically consider scenarios with failure of a single device or system, whereas an abnormal situation can involve simultaneous failure of multiple devices or systems. For such situations, perhaps a "What-If, HAZOPstructure can be used to brainstorm abnormal scenarios that should be considered, especially with respect to process alarms, emergency procedures, or emergency training drills. Using a "What-If" approach for an established plant that has been in operation for many years could be useful in highlighting events that have occurred but where lessons have not necessarily been learned, incorporated, or embedded into the operating procedures, culture, or practices. The HIRA facilitator must be familiar with the concept of ASM to direct the risk analysis team effectively during this "What-If" exercise.

## 5.3   PROCESS CONTROL SYSTEMS

Table 5.3 provides an overview of some of the strengths and weaknesses of tools that may be considered when developing strategies for monitoring, diagnosing, and predicting both process variances and abnormal process upsets.

## Table 5.3  Process Control Systems

| Common Tools and Methods | Strengths | Weaknesses |
|---|---|---|
| Monitoring of process trends | Algorithms can be programmed and left in place with minimal refreshing.<br><br>Automatic trends to provide warnings about potential process issues such as too fast, too slow, flat line, prior to alarms.<br><br>Can provide earlier indication if a process is slowly trending outside of safe limits.<br><br>Support for operator situational awareness and mental model. | The notifications that result from this added information need to be managed to prevent adding to alarm overload.<br><br>Poorly designed trends are not useful for diagnosing processes, especially those carrying higher risks. |
| Alarm Management | Prevent alarm flooding.<br><br>Focus on priority alarms.<br><br>Ensure visual and audible awareness of safe operating limits. | The process of reviewing all existing alarms and assigning them a priority can be a time-intensive exercise.<br><br>A formalized process must be established to ensure that alarms subjected to a rationalization process are not eliminated without due cause and MOC. |
| Fault detection and diagnosis | Can identify incipient problems that are outside the experience base of the operators<br><br>Use of such technology would likely have prevented some major disasters of the past. | Depending on the strategy taken, a current process model or a long history of process performance is required. Developing the logic could be very time consuming. |

**Table 5.3 Process Control Systems** *(cont.)*

| Common Tools and Methods | Strengths | Weaknesses |
|---|---|---|
| Big Data | Can be used to analyze large sets of data where the relationships between process variables, equipment, and operating modes are difficult to derive | May require additional capacity for data processing and expertise in configuring data to analyze. |

### 5.3.1    Process Trend Monitoring

The design of the control system should provide an interface so that the panel operator can readily observe patterns and trends of multiple variables simultaneously. This design exercise should be conducted based on consultation with the operating team and critical parameters that have been highlighted in previous studies (such as HAZOP/ LOPA). Procedures and training must include instructions on the correct use of such displays including those that should be checked periodically and others that can provide helpful information under certain abnormal conditions. Displays that provide an overview of the key operating trends, key parameters, and "big picture" of the entire process are especially important features of a well-designed control panel. For batch processes that frequently start up and shut down, trending can be designed to notify the control panel operator that the process may be beginning to experience an abnormal situation. Problems have occurred, particularly during non-routine or transient operations, where an instrument has gone out of range making it very difficult for operators to correctly diagnose the situation. The design of the system should enable instruments to stay in range under all expected circumstances.

Most modern control systems provide additional features, such as a rate of change alarm, which can be helpful in certain circumstances, although these must be set with care to avoid becoming nuisance alarms. The benefits of adding a well-designed rate of change alarm are shown in Example Incident 5.1.

---

### Example Incident 5.1 – Ring Drier Control

A process for making a solid resin for the paint industry included a section where a slurry of resin particles in water required dewatering. The first section involved a series of four, continuous centrifuges, arranged in a parallel formation. The centrifuge "cake" was dropped into an air-conveyed ring dryer where it was blown around a 40-foot (12m) diameter ring, comprising a 1-ft (0.3m) square section of ducting.

A relatively frequent problem was a blockage in a centrifuge that quickly led to liquid overload in the ring drier and the solid material slumping and caking on the inside of the ring. This required the feed to be stopped and the centrifuge to be unblocked. A large amount of cake deposited in the dryer would create an ignition risk unless it was fully removed before the process restarted. There had been several incidents of major fires over a period of ten years.

Following discussions with the operators, investigators established that if operators happened to spot the dryer outlet temperature dropping quickly, they could reduce the load and recover the situation quickly without blockage. The system had an adjustable low temperature alarm just below the operating value, but it was not always set correctly. Often the system was in and out of alarm and panel operators would lower the setting. The plant had worked this way for years, but a decision was finally made to fit new instrumentation to see if improvement could be made.

One of the key additions was a *rate of change alarm* that was added to the dryer outlet temperature. This fundamentally changed the operation of the process since operators could quickly identify the early signs of a centrifuge overload, reduce the feed rate, and recover the plant operation without slumping occurring in the dryer. The additional benefit was that without the slumping, there were fewer burnt particles in the resin product, which led to a major improvement in product quality.

---

### 5.3.2    Alarm Management

With the installation and advancements of DCS control systems, the significance of alarm management has become a major challenge for process control and process engineers to establish guidance on effectively recognizing and managing alarms during both normal and abnormal process situations. Design challenges such as alarm flooding, alarm suppression, alarm grouping, and alarm rationalization are commonplace. Process operators often encounter conditions such as alarm flooding, nuisance alarms, stale alarms, incorrectly prioritized alarms, and alarm overload. These concerns led industry to introduce standards that address the principles of alarm rationalization and alarm management, with respect to human factors and ergonomics affecting the front-line operators and supervisors. Section 4.2.5 includes a list of these published standards (EEMUA 1999/Revised 2013, NAMUR 2003, ANSI/ISA 2009/ Revised 2016, API 2010/Revised 2016, and IEC 2014).

These publications provide valuable guidance and recommended standardization for alarm management and are the tools and methods that are most often applied when designing or modifying alarms systems. Therefore, process control engineers and process engineers who provide input on alarm configurations must be familiar with these standards and guidelines to ensure the design addresses:

- Operating personnel are not overloaded with low priority or nuisance alarms.

- Control Panel Operators can smartly select an alarm graphic overview of each section of the process, and the overall unit, as required.

- Human factors are a major part of alarm management design and HMI.

In addition to these standards and guidelines, various companies have provided relevant tools and approaches to assist industry in the effective management of alarms. For example, Figure 5.1 (GCPS 2017) shows the hierarchy of process safety management and safeguard protection layers, where alarms are part of the system for preventing an unwanted event.

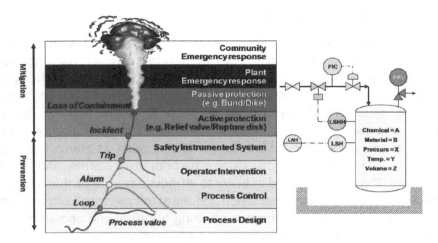

**Figure 5.1  Protection and Their Impact on the Process**

Unlike other safeguards or layers of protection, such as a pressure relief valve or safety instrumented system (SIS), the operator's response to an alarm relies on human intervention. There are numerous potential failure modes for operator response to an alarm including hardware, software, and human behavior. Failures in human behavior become more likely with poor alarm system design and performance (nuisance alarms, stale alarms, redundant alarms, and alarm floods). These failures are often improperly labeled as "operator error"; but are often more appropriately characterized as alarm management failures.

In summary, alarm management and control panel operators are critical layers of safeguard protection. However, when it comes to management of abnormal situations in the plant, traditional distributed control system (DCS) alarms are often not enough. Manufacturers have developed methods to apply boundary management tools that test and manage the process limits. To work effectively, however, a variety of structured and unstructured process input data must be aggregated in a common database for continual analysis. Only then can predictive analytics and root cause analysis be proactively applied to prevent processes from drifting into abnormal situations or unwanted events.

Example Incident 5.2 describes how multiple equipment failures occurring concurrently overwhelmed the operator consoles with alarms, confusing the control panel operators. Numerous opportunities to correct the situation were missed, resulting in the all-too-familiar Three Mile Island Reactor Core Meltdown and a massive evacuation effort. (CCPS 2008d).

---

**Example Incident 5.2 – Three Mile Island Reactor Core Meltdown, 1979**

On March 28, 1979, the pumps circulating water through the secondary loop of a pressurized water reactor failed at Three Mile Island Unit 2 located near Middletown, Pennsylvania. As a result, both the steam turbine driving the electric generator and the pressurized water reactor automatically shut down. Without the cooling water, the reactor began to build pressure, causing the pilot operated pressure relief valve (PORV) located in the primary reactor cooling loop to lift. The PORV should have closed when the pressure reduced to below its set point, but it did not. This allowed cooling water to continue flowing out the PORV, causing the reactor core to overheat and ultimately experience a meltdown. Approximately 144,000 people were evacuated from the area for 5 days; however, no one was injured or suffered irreversible effects.

When the pump failed, it was difficult for the control room operators to determine what happened as instrumentation was telling a confusing story. No instrumentation was installed to indicate the level of coolant surrounding the core, so operators judged the level by using the level shown in the pressurizer vessel on which the PORV is installed.

This led to a misconception that the reactor core was flooded. Since they saw no need for emergency cooling water, they turned off the pumps that would have provided additional water. A backup emergency cooling water system had also failed, as valves had not been realigned following a recent test of the system. Additionally, there was no indication that the PORV remained in the open position.

---

**Example Incident 5.2 – Three Mile Island Meltdown (*cont.*)**

In the words of one of the control room operators, the console "lit up like a Christmas Tree." Despite the enhanced design and multiple layers of protection, there were equipment and human failures that prevented returning the reactor to a safe condition.

Lessons learned in relation to abnormal situation management:

- Plant Operator Training and Understanding: Although the design incorporated multiple layers of cooling protection, operator training failed to address this particular scenario and the abnormal situation of a stuck open relief valve was not recognized.

- Process Monitoring: Instrumentation that could have helped diagnose the reactor water level was not included in the design. When the upset occurred, the control and alarm panel was flooded with lights and alarms.

- Procedures: Failed work procedure for critical backup emergency equipment as the pump system was not returned to the active position following recent testing.

---

### 5.3.3    Big Data

The collection of large amounts of data from processes is an ongoing activity, for example via historians and other databases. It is only relatively recently, with the continued improvement in computer processing power, that tools have been developed to make valuable use of large data sets, or "big data". For example, big data can be analyzed to reveal patterns, trends, associations and, using advanced tools the data can be extrapolated to make predictions about future performance. This is particularly relevant to abnormal situations and may involve the data of an entire refinery operation with respect to interactions from raw feeds, all the process units and equipment, utility supply units, and the finished product distribution.

There are an increasing number of commercially available visual analytics tools that greatly simplify the task of reviewing historical time series and event data, helping the user to identify patterns in the data and search for prior occurrences of these patterns. These visualization tools can be used to help the operations and engineering teams track abnormal situations, investigate incidents, and monitor plant operations.

More digital devices for data collection are rapidly becoming available, including wireless equipment that can be relatively inexpensive to install. These can provide additional information on the conditions of a process and the associated mechanical equipment, including such attributes as vibration and temperature monitoring, noise levels, pipe wall thickness and gas detection equipment.

The effective use of the data provided by the process and the data analysis tools provides information that can result in many improvements including product quality, operating efficiency, equipment reliability, diagnostics, establishment of more appropriate alarm limits, and other key performance indicators, as further discussed in 5.3.2 and 5.3.4.

### 5.3.4    Advanced Diagnostics and Artificial Intelligence

Advances in smart devices and data collection, computer processing power, analytics and machine learning have opened up new approaches to monitoring and predicting the performance of a process. Advanced tools, such as anomaly detection machine learning methods, are designed to be forward looking rather than providing a historic overview, such as those that were formerly provided by historians and process alarms. These methods primarily rely on a data model, calibrated to plant history, to provide advance indication of equipment deterioration. Predictive analytics and artificial intelligence methods do have limits. Having more data does not always lead to more valuable insights. Some common situations that can limit the effectiveness of these systems include:

- Some measurements, such as controller outputs or setpoints, are not be stored in long-term history.

- Some data is not recorded at a high enough frequency for the required analysis or there may be too much data compression applied.

- Some plant data does not have sufficient variability to establish the necessary model relationships. For example, a plant that operates in a steady state condition with minimal change in process variables provides the worst type of data for developing a data analytics model.

Digital twins that can diagnose, predict, and forecast performance of a process and the equipment provide a step forward from the more traditional predictive maintenance techniques, which usually rely on relatively limited data. By combining process data, condition-monitoring data, historic performance and artificial intelligence, advance warning of process problems and equipment failure becomes much more reliable. With fewer equipment breakdowns comes a reduction in the number of abnormal situations and improvements in safety as well as operating and maintenance costs generally follow.

## 5.4   POLICIES AND ADMINISTRATIVE PROCEDURES

Table 5.4 includes some of the policies and administrative procedures that facilities have established and communicated as minimum expectations of personnel with respect to decision-making and information sharing.

## Table 5.4 Policies and Administrative Procedures

| Common Tools and Methods | Strengths | Weaknesses |
|---|---|---|
| Organizational chain of command, hierarchy | Defines responsibilities and authorities | Limitations of authority can create problems in abnormal situations when critical decisions must be made quickly.<br><br>At large facilities, area management can devolve into 'kingdoms' that result in inconsistent standards across the site. |
| Communications between shifts – verbal, written logbooks and electronic | Provides seamless link between shifts so that transient conditions are managed. | In practice, the quality of shift change communications is highly variable, and requires supervisor monitoring. |
| Auditing of conformance to policies and administrative procedures. | Can detect gradual degradation of systems and behaviors that may not be apparent to people working day-to-day. Can also detect specific faults that would aggravate an abnormal situation. | Findings are a snapshot in time so that failings that occur between audits may be present for a year or more before being detected. |
| Process Metrics | Key Performance Indicators (KPIs) can be established to monitor the alarm system data as well as process parameters.<br><br>Excellent for checking current process conditions against target parameters.<br><br>Similar metrics may apply across multiple processes. | May provide a superficial view of "symptoms" rather than underlying faults. |

## 5.4.1    Expectations of Policies and Administrative Procedures

Clear and accepted policies and administrative procedures are essential to establishing the minimum expectations of personnel. For example, are the policies clear to all personnel regarding their authority to make timely decisions?  Note this includes the "Stop Work Authority", which allows any personnel to request a halt in procedures or operations if there are safety concerns. Are organizational responsibilities documented?  Are communication requirements between operating shifts, between operating teams and maintenance, and between operating teams and leadership written down and followed?  Formal, written policies and procedures are recommended over depending only on guidelines. Often, guidelines can imply that the expectation of conformance to the regulations is optional or that they are merely suggestions.

## 5.4.2    The Relationship of Policies to Abnormal Situation Management

Policies and administrative procedures evolve over time as a company or plant site matures. Most companies now have established policy manuals that cover safety and environmental procedures, as well as onsite and offsite emergency response procedures. Management guidelines are then written to manage changes to these policies and procedures. Policies also establish a working culture, for better or worse. Cultural issues are a recurring theme in some of the example incidents in Chapter 3. A positive working culture indicates:

- An environment that respects and supports each team member and decisions that are made, without the benefit of hindsight.
- Admitting, and sharing/learning from, mistakes.
- Proactive seeking of learnings from others.
- Open and structured communications between operation teams and others.

It is not always easy to establish an effective relationship between teams. Plant operating teams/shifts tend to build autonomy among themselves especially in the presence of a strong or dominant team leader. While this can promote team pride and build intra-team relationships, from a more holistic view, it can also be detrimental to safely operating a chemical process if ongoing plant process conditions and abnormal situation issues are not shared with other groups.

A key element is communication, not only within but also between shift teams so that any abnormal situations can be addressed and safely managed. Some traditional tools and methods to address these types of communications include:

- Traditional shift-to-shift or shift hand-over communications conducted verbally and/or written.

- Adding descriptive notes and messages in an operations logbook.

Neither approach can assure that the communications are fully understood or whether they are truly reflective of a possible abnormal situation. This is true with modern 12-hour shift patterns where there are frequent long periods of absence from the site, which can make it cumbersome or ineffective to wade through many days or weeks of logbooks to determine the status of a process. More structured communication tools are recommended, including standardized handover log sheets, checklists, and a record of longer-term issues. Some companies use a system whereby technicians and operators must sign a communication page, to ensure that key long-term changes or issues have been communicated and understood. Much of this can be achieved with computer software although such a system should be used to support, rather than replace, well-designed and properly followed handover procedures.

Documents and files, whether physical or within a computer system, can be structured to summarize both normal and abnormal conditions encountered during the shift. They can also provide a valuable source of information to identify process areas of concern for further investigation or for process improvement discussion with a team of process experts. Some facilities have a policy of specifying the point in time at which handover to the new shift team is complete to emphasize that they now have ownership of the operating unit and are responsible for its condition going forward.

Some communication tools are preferred over others, although no matter which tools are in place, they need to be understood and used effectively by all. Human factors form a major part of effective communication and handover, and it is recommended that staff who use such systems be involved in their design and setup.

The BP Texas City 2005 Example Incident 3.3 already discussed in Chapter 3 of this book is an example of a lack of a clear line of responsibility and effective communications between operation teams.

Example Incident 5.3 illustrates how poor communication between operating teams and plant supervision contributed to a serious event at a chemical facility.

---

### Example Incident 5.3 – Hydrogen-in-Chlorine Explosion

Chlor-Alkali plants produce chlorine, hydrogen, and sodium hydroxide. One very well referenced risk known about hydrogen-in-chlorine explosions is that if the concentration reaches a range from 4-8%, the potential for an explosion exists. The normal concentration of H2 in the chlorine gas stream is in the low parts per million (ppm).

This event occurred during a plant startup, typically the operating stage with the highest risk for chlor-alkali plants. The plant had been shut down for weeks for a major turnaround. New equipment including new electrolyzers and instruments and controls were installed. During the startup, the plant operating teams were experiencing abnormal conditions. Troubleshooting was done across shifts. Plant leadership was aware of this but encouraged the operating teams to continue trying to work through the problems. The unit supervisor and manager resisted shutting down the plant, so the operating teams did not feel that they were empowered to shut down the unit.

After many hours of trying to line the process out, a hydrogen-in-chlorine explosion occurred in the chlorine header. The explosion did not cause any injuries as no personnel were allowed in the process area during startups; however, the explosion led to a small release of chlorine gas, extensive equipment damage, and serious business interruption.

---

**Example Incident 5.3 – Hydrogen-in-Chlorine Explosion** *(cont.)*

The incident investigation found several contributing factors:

- The new pressure transmitters were wired backwards.
- No PSSR was conducted to check the pressure controls before startup.
- Leadership was under pressure to start up the plant as the schedule was past due.
- Communication between plant leadership and operating teams was strained due to several issues encountered during the shutdown.

Lessons learned in relation to abnormal situation management:

- Organizational Chain of Command: During this stressful abnormal situation, the unit leaders overrode the chain of command and empowerment of personnel, who did not consider they had "Stop Work Authority".
- Management of Change- Pre-Startup Safety Review:  Although the MOC was conducted for the changes, the PSSR was not conducted because startup was overdue.
- Learning from Incidents: History across the chlor-alkali industry has included many hydrogen-in-chlorine explosions. The highest risk is during startups. This was not considered during this plant's startup.

---

### 5.4.3    Process Metrics

Metrics are an important consideration of many business models. Metrics have been created for business goals, quality, safety, environment, security, training, and mechanical integrity. Metrics are also highly relevant in the management of abnormal situations. It is worth noting that CCPS considers measurements and metrics of such high importance that they are included as one of the four elements in the RBPS pillar of *Learning from Experience*. (CCPS/RBPS 2007a). CCPS has also published *Guidelines for Integrating Management Systems* and *Metrics to Improve Process Safety Performance* (CCPS 2016c). Both books provide excellent guidance on establishing and maintaining an effective metrics process. Metrics are discussed in Chapter 6 Section 6.2.

## 5.5    OPERATING PROCEDURES

Table 5.5 lists two tools for evaluating procedures to assess if they are current, correct, and complete. Procedures are often considered the first tier of human response safeguards to prevent an unwanted process situation, however when the procedures are outdated, missing key information, or incorrect, the likelihood of a process upset occurring is increased.

### Table 5.5  Techniques for Reviewing Operating Procedures

| Common Tools and Methods | Strengths | Weaknesses |
|---|---|---|
| Transient Operation HAZOP | Focuses on hazards during transient operations that history shows are more likely than hazards during normal operation. | Not all abnormal situations can be predicted, therefore, all-purpose emergency protocols will always be needed. |
| Procedure HAZOP | Like a standard HAZOP, this type of study provides structure to a review of written procedures. | If the procedures are not already in good condition (see bullets in Section 3.4.1), this will be time-consuming. |

### 5.5.1   Standard Operating Procedures

Standard Operating Procedures (SOPs) are one of the 20 fundamental elements in the CCPS' *Guidelines for Risk Based Process Safety* (CCPS 2007a) and are required by many government agencies. Therefore, by expectation, SOPs are one of the many established process safety tools that most companies already have in place to prevent and if necessary, mitigate incidents. The scope of SOPs such as startup, normal operation, and shutdown can extend beyond the standard simple wording such as "start feed flow, heat up, open valves, monitor level" content that appears in most written procedures.

For example, Example Incident 3.8 – Distillation Column Startup from Chapter 3, Section 3.4.2.1 illustrates the need for a better initial commissioning procedure in addition to a process startup procedure.

SOPs should be structured to include safe operating range, provide warnings against exceeding the safe operating limits, and include recommended steps to bring the process back into safe status. However, since procedures are called "Standard Operating" Procedures, they often seem routine and do not specifically address the Abnormal Operating Situations.

In order to address this, the ASM Consortium® conducted a research study to investigate procedure execution failures in abnormal situations (Bullemer, Kiff & Tharanathan 2010b). The study team examined data related to procedural operation failures across a data set from 32 public and private incident reports. The main finding from this investigation was that the majority of the procedural operations failures (57%) across these 32 incident reports involved execution failures in abnormal situations.

The analysis of the top causes of procedure execution failures found:

- The most common failure was associated with lack of knowledge about appropriate responses to the occurrence of an abnormal situation while executing a procedure.

- The second most common failure was the failure to detect the presence of an abnormal equipment or process mode while executing a procedure.

- The third most common failure was the lack of understanding the impact or effect of a procedural action or failure to execute a procedural action.

In total, these top three failures account for 35 of the 40 (87.5%) procedural execution failures under abnormal situations. [ASM® Guideline, *Effective Procedural Practices*] *(ASM® Consortium - Bullemer, Hajdukiewicz & Burns 2010).

Typical tools and methods for evaluating the content and accuracy of process procedures are similar to HIRA tools: HAZOP, What-Ifs, and Checklists.

An effective approach is to conduct a procedure observation audit that involves an observer following along with the procedure as the operator conducts the work. The observer ensures that the steps are followed in order and highlights where any steps are missed, or identifies confusing items. Another approach is to ask a newer operator to do this task following the steps in the procedure (highlighting any missing or confusing items that are encountered).

Ensuring that operator technicians follow the most current approved procedure can be a challenge. Printed procedures can quickly become uncontrolled copies that are outdated or damaged in use (e.g., spills, tears, missing pages). Electronic copies of procedures can be difficult to access quickly in an emergency, especially if a power or computer network outage occurs. Thus, a suitable balance has to be struck and procedure control management practices must be established, communicated to personnel, and adhered to ensure that up to date procedures are followed.

## 5.5.2   Emergency Procedures

Responding to emergency conditions requires planning and training. Identifying the normally anticipated types of initiating causes for emergencies such as chemical leaks, fires, loss of a critical utility supply, on-site transportation events, or natural weather events are often already in place across most of the chemical industry. Emergency procedures for unusual, infrequent, or near impossible events, however, are not as easy to develop. Therefore, identifying in advance possible abnormal situations that may require an emergency response should start with the HIRA studies of the process plant and associated utility systems. The studies should be led by a trained facilitator who has knowledge of the concept of abnormal situations. Once the possible list of scenarios has been created, the site's risk analysis or process safety experts may want to build fault trees or Bow Ties, or to use similar tools to determine if the scenarios have adequate preventive safeguards. Development of emergency procedures may also be required, as one of the safety layers of protection for the facility.

### 5.5.3   Transient Operation Procedures

Operating procedures during transient phases of an operation (startup, shutdown, slowdown, pause, special trials, manually executed tasks, or emergencies), have often been identified as a contributing cause to many major incidents. Chemical processes designed to operate continuously will often experience many more incidents per unit time during a transient phase than in steady state operation. Likewise, for incidents associated with flying such as during takeoffs, when landing, or in extreme weather conditions, such a possibility is highest at those times. Ensuring that accurate, concise, and easy-to-execute procedures have been developed and are in place for these operational phases is essential to safely managing many of the abnormal situations that may arise. Furthermore, deficiencies in training and incomplete or inaccurate procedures are often cited as root or contributing causes for incidents.

HIRA tools such as HAZOPs and What-Ifs are typically utilized to assess the need for and scope of the Standard Operating Procedures. However, for the transient phase operations, a more specific tool such as Transient Operation HAZOP (TOH) can be used. A common element in transient operations is the requirement for increased human interaction with the process. Often the operator and procedural controls are the key layers of protection for preventing an incident.

Reduced operator experience — because of retirements, longer turnaround intervals, and more reliable units — frequently results in more reliance on procedures as a source of information and a critical layer of protection against process hazards. The TOH differs from a conventional HAZOP since it focuses on operational tasks and procedural controls, which are believed to be critical, and which often are not discussed in detail during a more traditionally focused HAZOP.

The TOH process centers on identification of required unit-specific activities (tasks) with a potential for an acute loss of containment and an in-depth review of the procedural controls necessary for safe and successful completion of those tasks. Timely identification of hazards, adequacy of procedural and design controls to ensure correct sequencing, early feedback of potential errors, clarity and completeness of transient operations all are carefully assessed. The technique uses a combination of knowledge and experience of a cross-functional team, guide words and reference lists to drive a disciplined approach to identify enhancements for procedural and design-related issues. Further details on TOH are available in a presentation to the European Conference on Process Safety: *A HAZOP Methodology for Transient Operations* (Ostrowski & Hertoghe 2019). CCPS has published several books associated with procedures and hazard analysis. Three of these are *Guidelines for Risk Based Process Safety, Guidelines for Writing Effective Operating and Maintenance Procedures*, and *Guidelines for Hazard Evaluation Procedures* (CCPS 2007a, 1996, 2008b).

### 5.5.4    Preparing Written Procedures

Much has been documented about the subject of preparing effective written procedures. Some key principles include:

- Involving both the operations and engineering staff in the development process.

- Using a step-by-step approach rather than lengthy paragraphs.

- Providing warnings of potential hazards before, not after, the step in which the hazard is present.

- Using graphics to provide more information, and visual stimulation.

- Use of signoff boxes to increase likelihood of operator engagement with each step.

- Performing a field walkthrough to validate that the step sequence is achievable and understandable by the least-experienced staff member who will be performing the work.

The ASM® Guideline, *Effective Procedural Practices* (ASM® Consortium - Bullemer, Hajdukiewicz & Burns 2010), provides recommendations for the effective development and use of procedural practices. The guidelines can be used to assess the quality of a company's procedural practices from the perspective of their potential impact on abnormal situation management. The focus of the guidelines is for procedures that operators use while operating, but the guidelines are generally applicable to all types of procedures used on site.

## 5.6    TRAINING AND DRILLS

Training of personnel to understand and safely execute their work assignments whether they are control panel operators, field operators, maintenance personnel, or support personnel is a fundamental element of risk-based process safety. Table 5.6 provides an overview of typical training tools that can be adapted for training on abnormal situations.

## Table 5.6  Training and Drills

| Common Tools and Methods | Stre12ngths | Weaknesses |
|---|---|---|
| Modular training tool | Convenient for fitting into operators' work schedules. Limited scope makes retention more likely. | Periodic refresher training is likely to be needed. Training may ultimately be wasted if the worker leaves the company or gets an unrelated assignment. |
| Unit-Level Emergency Drills | Repetitive drilling is possible due to smaller scale than a site drill. | In some instances, a drill or test has actually <u>caused</u> an event (e.g., Chernobyl) |
| Site- or Offsite-Level Emergency Drills | Checks for communications between all functions. Post-drill critiques can identify response steps to improve. | Extensive planning is required to involve all functions, especially off-site responders. |
| Evacuation Drills | Can be used to ensure primary muster locations and backup locations are safe. Good for onsite construction personnel to participate. | Difficult to ensure everyone has an opportunity to participate, especially if there are multiple shifts at a plant. |
| Process Control Simulation | Excellent for practicing actual process control steps. Can be used to simulate abnormal process situations. | Process simulator must be maintained to match the plant control system exactly. |
| Field Operator training to use their senses (hear, see, smell) | Important training to recognize abnormal conditions in the field. | Frequent refresher training is often needed as memory of abnormal events and incident details can quickly fade. |

The skill of the control panel and field operators is directly related to the effectiveness of their response to most abnormal situations. An example is provided by the *Example Incident – Tower Flooding* that was presented in Section 3.4.2.2 (see Example Incident 3.10, Ch. 3). In that incident, most if not all the appropriate information was available to the control panel operator, but the instinctive response only served to aggravate the situation. Several steps might have been taken in advance to prevent the problem, such as:

- Educating the control panel operators on the principles of distillation columns, including flooding of this type.

- Providing additional information to the operator – for example, a pressure differential reading between the top and bottom of the tower that could indicate too much vapor flow in the tower.

- Using process simulators to enhance operator training.

When developing the training scope and approach, introduction of abnormal situation management can be added as a training module. The concept of ASM® should be introduced as well as demonstrated with a few examples inherent to the chemical process that is familiar to the personnel. Then ASM®, in addition to issues identified from HIRAs, can be interwoven into frequently used and often recognized training tools and methods such as:

- Formal training manuals that address fundamentals such as process engineering parameters, basic process operating design, chemistry, chemical, and fire risks.

- Tabletop training exercises with desired responses on abnormal situations and scenarios.

- Emergency situations and response drills.

- Alarm response training.

- Process simulation of the process control systems.

- E-module training.

- One-on-one training.

Operator training simulators can be an important tool in reinforcing a mental model and building "muscle memory" responses to key fast acting issues. Emerging technologies are allowing drills to be developed with a more immersive experience for staff including virtual reality "gaming" type simulations. These have the added benefit of simulating some of the stress and key time-related aspects of incidents that are not present in most paper-based drills

Another focus of training is for field operators to stay aware of their surroundings by using their natural senses such as sight, hearing, or smell. This can be important in detecting and recognizing abnormal conditions related to field equipment. Some examples are:

- Equipment vibrations
- Cavitating pumps
- Missing plugs or caps
- Unexpected ice or frost on equipment or piping
- Smell of burning rubber or insulation
- Squeaking belts
- Unusual odors
- Open electrical boxes or fittings
- Corrosion
- Leaks and drips
- Layers of combustible dust
- Missing or damaged pipe hangers or supports

These are just a few examples of abnormal situations that should be investigated quickly. Other examples can be created and customized for the specific process that personnel are being trained to operate.

Note: To read more about training and knowledge, please refer to Chapter 4, Section 4.3.

## 5.7    ERGONOMICS AND OTHER HUMAN FACTORS

Table 5.7 addresses several key factors and working conditions that can negatively affect the performance of plant personnel. Control room layout, process control console configuration, environmental conditions in the control room, and alarm design are just a few of the most important conditions. Although control room personnel will readily adapt to their working conditions, the impact on their performance and response to emergency situations may not be apparent until an actual situation occurs.

### Table 5.7  Ergonomics and Other Human Factors

| Common Tools and Methods | Strengths | Weaknesses |
|---|---|---|
| Process Control System Graphics and Displays | Control system displays provide operators the overall status of the process.<br><br>Graphics can be designed to prioritize highest risk areas.<br><br>Graphics can support troubleshooting and effective response during abnormal situations. | Poorly designed displays and graphics may cause confusion during abnormal situations. |
| Ergonomic Assessment | Improve front line operator working conditions.<br><br>Reduce stress during abnormal or emergency situations. | May be costly to implement improvements to older control rooms. |

Control system display screens and process control panels are often where abnormal situations are first recognized. Resolving the situation usually involves a human response defined by an organization and its policies. Example Incident 3.2– *Bhopal* from Section 3.1.1 is a classic example of this (although in that case the chemical release was identified through smell and visual observation of a cloud). This section includes key components of the process control system that can be a tool for recognizing an abnormal process condition or a detriment if it is poorly designed.

### 5.7.1    HMI (Human Machine Interface) System

The primary tools available to the control panel operators to help them to "recognize and understand" an abnormal situation are the instruments, control panels, console graphics, and process alarms that provide a status of the operating situation to the operators at all points in time. While process control systems have advanced greatly over the past few decades from analog to digital presentations, these "advancements" do not always have the desired effect. Across a broad range of processes, such as chemical, mechanical, medical, and transportation, the primary information tool for front-line personnel will be the process control system and associated control screens and graphics. Therefore, the overall design, layout, and content of the console screens and graphics for each process step or task should be fully considered and discussed with respect to the human interface.

Figure 5.2 shows the mental and physical process that the operator performs in order to build up an accurate mental model of what is happening in the process to help correctly handle abnormal situations (Downes 2017). A well-designed HMI simplifies many of these activities.

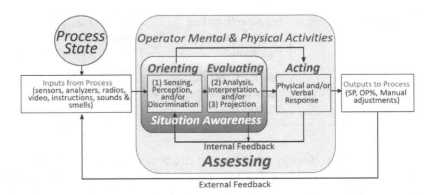

**Figure 5.2  Model of Mental and Physical Processes in Process Control**

Some general aspects of the Human Machine Interface (HMI) to consider:

- Are indicators for operation critical and safety critical process control parameters and process conditions included, and if so, are they visually correct and readily observable?

- Are the graphics overwhelming due to an excessive number of low priority parameters and alarms?

- Is an alarm summary list always displayed to the operator for quick review?

- Is there consistency between the control screen graphics with respect to layout, symbols, and colors?

- Is there good resolution and minimal glare?

- Are the procedure steps for manually initiated control panel actions proven and straightforward to bring the process back to a safe state?

- Is there an overview screen (level 1 graphic) available for rapid access from all other screens that provides key overall operating parameters in the event of an abnormal situation?

- Is it simple for the operator to drill down from the level 1 graphic into specific areas of interest?

- Can trends on critical process parameters be created and observed?

- Can historical operation reports be generated for review?

The considerable variety of HMI designs can lead to human factors issues. For example, in some designs, red indicates "off" or "closed" and green means "on" or "open." Other designs may adopt exactly the opposite convention, where red is "danger", i.e., "on" or "open". Other systems use grey and black (similar to a P&ID). Control stations are continuing to advance with improvements in resolution, video display, and human interface such as a touch screen. Some standardization has taken place more recently, as detailed in 5.3.2.

Operator interface design has been an important focus of the work conducted by the Abnormal Situation Management® Consortium. Following site practice assessments and fundamental human factors research, the guideline, *Effective Console Operator HMI Design* (Bullemer & Reising 2015) provides details of ASM® Consortium recommendations for designing information displays and devices for console operator workstations. The primary audience for this document is individuals who establish or assist in establishing company standards, style guides or evaluate vendor capabilities for information displays in console operator workstations. The document contains:

- Specific guidelines for achieving compliance with ASM® Consortium recommended practices, including priorities to indicate minimum requirements for compliance and requirements to achieve a best practice.

- A description of "How it Works" to enable assessment of whether the management system used to develop information displays and devices in the console workstation provides appropriate and sufficient functionality to enable operators to effectively manage abnormal situations.

A second Abnormal Situation Management® Consortium guideline, *Effective Change Management Practices in HMI Development* (Bullemer & Reising 2021) provides recommendations on how to upgrade HMI displays to ASM style displays in order to achieve operator acceptance.

The benefits of deploying ASM style graphics were evaluated during an ASM study (Reising & Bullemer 2009). Over a sequence of simulated abnormal situations, significant and consistent reductions in response times (on average 41% better) were observed for operators using ASM style HMI interfaces. Furthermore, the operators were more proactive, detecting more events before an alarm and more effective in correctly responding to those alarms.

## 5.7.2    Control Room Ergonomics/ Human Factor Assessment

The surroundings in which people work can affect not only their routine performance, but also be a factor in their quick and safe response to an abnormal situation or an emergency process upset. An Ergonomic Assessment is the tool most often applied to review humans and the impact of their physical surroundings. When an ergonomic assessment is conducted, several workplace design factors that affect human performance are typically considered:

- **Control Room Lighting** - understand current best practices in control room lighting, variability across control rooms and within the same control room, variability across shifts and the time of day, ability of control panel operators to adjust the lighting to suit personal preferences and changing conditions.

- **Climate Comfort** - review best practices and methods for measuring work environment climate comfort (such as indoor air quality, temperature, lighting, background noise, and odors)

- **Control Room Layout**—access and egress, lines of sight, physical separation walls, workstation spacing, temporary barriers, and obstacles.

Ergonomic design of control rooms is a serious aspect of providing plant personnel an environment for achieving high performance. This is of such importance that it is addressed by International Organization for Standardization ISO 11064 (ISO 2000a, 2000b, 2002, 2004a, 2004b, 2005, 2006). This ISO standard covers design principles, arrangements and layout, workstations, displays, controls, interactions, temperature, lighting, acoustics, ventilation, and evaluation. Other guidance is provided by Engineers Equipment and Materials User Association (EEMUA 2002) and the Abnormal Situation Management® Consortium, *Effective Operations Practices*, Section D.6 (Bullemer 2020). This topic as well as the issue of good ergonomic design in the field is also discussed in the CCPS book *Human Factors Methods for Improving Performance in the Process Industry* (CCPS 2007c). The CCPS book specifically states that the biggest challenge in process control is management of abnormal situations.

Many other aspects of human performance are relevant to the effective management of abnormal situations. The way a team performs together can be a key factor in determining whether the outcome was successful.

### 5.7.3   Crew Resource Management

The aviation industry had problems in this area and developed the term "Crew Resource Management" following a National Transportation

Safety Board (NTSB) investigation into a DC-8 aircraft that crash-landed in 1978 (NTSB Report), as illustrated in Example Incident 5.4.

---

### Example Incident 5.4 – Flight 173 DC-8 Crash in Portland, 1978

On December 28, 1978, United Airlines Flight 173, a McDonnell-Douglas DC-8 took off from JFK Airport in New York City bound for Portland, Oregon, with a stop-off at Denver, Colorado. It departed Denver at about 14:47 hrs with 189 persons on board and was due to arrive at Portland at 17:13 hrs. The flight was delayed on approach to Portland while the crew investigated problems with the landing gear. It ran out of fuel and crashed into woodland at about 18:15 hrs on approach to Portland resulting in 10 fatalities, including the flight engineer and a flight attendant. There were also 23 serious injuries.

According to the National Transportation Safety Board (NTSB) report [Reference 1]:

> *"The probable cause of the accident was the failure of the captain to monitor properly the aircraft's fuel state and to properly respond to the low fuel state and the crewmember's advisories regarding fuel state.　This resulted in fuel exhaustion to all engines. His inattention resulted from preoccupation with a landing gear malfunction and preparations for a possible landing emergency."*

The investigation found a fault in the landing gear lowering mechanism. The gear was down safe, although a microswitch was damaged so the crew did not get the normal indication.

The flight engineer had expressed concern about the fuel at 17:50:47, stating: *"Not enough. Fifteen minutes is gonna really run us low on fuel here"*. At 18:07:06, the first of the four engines flamed out; the second at 18:13:21, and Mayday was declared at 18:13:50.

> **Example Incident 5.4 – Flight 173 DC-8 Crash in Portland, 1978**
> *(cont.)*
>
> The NTSB considered that the accident was an example of a recurring problem:
>
> > *"... A breakdown in cockpit management and teamwork during a situation involving malfunctions of aircraft systems in flight."*
>
> The report continued:
>
> > *Admittedly, the stature of a captain and his management style may exert subtle pressure on his crew to conform to his way of thinking. It may hinder interaction and adequate monitoring and force another crewmember to yield his right to express an opinion. The first officer's main responsibility is to monitor the captain. In particular, he provides feedback for the captain. If the captain infers from the first officer's actions or inactions that his judgment is correct, the captain could receive reinforcement for an error or poor judgment.*
>
> The final recommendation in the NTSB report was as follows:
>
> > *"Issue an operations bulletin to all air carrier operations inspectors directing them to urge their assigned operators to ensure that their flight crews are indoctrinated in principles of flightdeck resource management, with particular emphasis on the merits of participative management for captains and assertiveness training for other cockpit crew members."*

The investigation led to the development of assessment and training on Crew Resource Management (CRM). Today, CRM has evolved to cover many issues that are highly relevant to the management of abnormal situations. Outside the aviation industry, it is sometimes called Team Resource Management (TRM) or Non-Technical Skills (NTS, or NOTECHS). It can be defined as *"the cognitive, social and personal resource skills that complement technical skills, and contribute to safe and efficient task performance"* (Flin et al 2003). It primarily focuses on leadership techniques and effective management of resources, but also concerns the cognitive skills that are required to gain and maintain situation (or situational) awareness, particularly in stressful situations. The International Association of Oil and Gas Producers (IOGP) produced a guide *"Crew Resource Management for Well Operations Teams"* (IOGP 2020) that conducted a literature search and survey then listed key NTS categories and elements for wells operating personnel. These are also relevant to other hazardous industries and are as follows:

**Table 5.8  Non-Technical Skills, Categories and Elements**

| Category | Elements |
|---|---|
| Situation Awareness | • Gathering information<br>• Understanding information and risk status<br>• Anticipating future state/developments |
| Decision Making | • Identifying and assessing options<br>• Selecting an option/communicating it<br>• Implementing and reviewing decisions |
| Communication | • Briefing and giving feedback<br>• Listening<br>• Asking questions<br>• Being assertive |
| Team Work | • Understanding own role with the team<br>• Coordinating tasks with team members/other shift<br>• Considering and helping others<br>• Resolving conflicts |
| Leadership | • Planning and directing<br>• Maintaining standards<br>• Supporting team members |
| Performance shaping factors - stress and fatigue | • Identifying signs of stress and fatigue<br>• Coping with effects of stress and fatigue |

Many of these categories are appropriate across the process industries.

Practical measures have been developed that can assess both the process of acquiring situational awareness (SA) and the product of situational awareness. Improvement of situational awareness seems to focus on two main strategies, either the design of the system interface to encourage better sampling and reduce the cognitive workload or training in situational awareness at the individual and team levels.

CRM assessment and training can therefore be an effective tool to help in the management of abnormal situations, and not just in the aviation sector. It has been adopted in other areas including medical, shipping, nuclear, as well as the oil and gas sectors. The assessment uses a behavior rating system based on a defined set of skills, with their component elements and associated examples of desirable and undesirable behaviors.  Suggested content for a training syllabus for CRM can be found in the IOGP guide (IOGP 2020) and in the Energy Institute Guide (Energy Institute 2014).

For the chemical and process industries, the shift teams could undergo a similar assessment to ensure they work together effectively, especially during an abnormal situation.


## 5.8    LEARNING FROM ABNORMAL SITUATION INCIDENTS

CCPS has included *Learn from Experience* as a pillar for Risk Based Process Safety and listed *Incident Investigation* as one of the primary elements under Learn from Experience. Table 5.9 references some tools that are frequently applied to learning from incidents and can be applied to abnormal situation incident prevention.

## Table 5.9  Learning from Abnormal Situation Incidents

| Common Tools and Methods | Strengths | Weaknesses |
|---|---|---|
| Near-Miss Analyses | A review of the learning from previous near-misses can be used to help prevent the occurrence of more significant incidents in the future. | Rules should be established to define a "near-miss" within an organization so corrective actions can be consistently applied. |
| Previous incidents (internal and across industry) | May be used to identify abnormal situations that could cause an incident. | Requires maintaining an available list of incidents to share with personnel, unless learning is embedded into procedures. |
| Fault Trees or Bow Ties | Can be used graphically to depict paths for abnormal incidents to occur and safeguards to prevent. | Personnel must be trained to construct the graphical model to reflect correct content and results. |
| Process Metrics | Can be used as Leading Indicator of process upsets or non-acceptable situations. | Metrics should be meaningful and not a target disguising underlying process safety issues. |

Most companies and facilities have a procedure for investigating incidents and accidents. However, an abnormal situation may have been prevented from becoming a major event because it was stopped by another barrier (procedural or physical). Unless the company has a culture and procedure for identifying and investigating "near-misses", including getting to the root cause and not simply concluding it was due to "human error", a learning opportunity may be lost. The facility may not be so lucky next time when that other barrier might not be performing as well.

Where possible, learning should be embedded into the operation through hardware, software, or procedural changes. It is not always easy to identify the occurrence of an abnormal event, but this can be helped by having a good reporting culture, perhaps assisted by a system of automated metrics reporting as described in Chapter 6.

The tools for investigating and learning from incidents are available in recognized and available publications. Therefore, rather than discuss them in this book, some suggested CCPS published reference books on Incident Investigations, Metrics, and Bow Ties are:

- *Guidelines for Investigating Process Safety Incidents*, 3rd edition 2019 (CCPS 2019)

- *Guidelines for Integrating Management Systems and Metrics to Improve Process Safety Performance* (CCPS 2016c)

- *Bow Ties in Risk Management* (CCPS 2018a)

## 5.9    CHANGE MANAGEMENT

Changes to processes and organizations occur for several reasons such as equipment upgrades or failures, process optimizations, design changes, new products, and business impacts. These changes may be very positive when managed thoroughly but may have just the opposite result if not actively managed and fully evaluated. Table 5.10 references the management of change procedure for processes as well as organizations.

## Table 5.10 Change Management

| Common Tools and Methods | Strengths | Weaknesses |
|---|---|---|
| Management of Change Procedure | Structured MOC procedure can identify areas of risk created by a change. | Must have a culture where changes are identified/recognized so that they can be raised up for review. |
| Management of Organizational Change | Can identify an issue within the organization or created by the change. | Requires a concise scope of job function responsibilities. |
| Pre-Startup Safety Review | Provides a check that the change was implemented as approved and that personnel are aware of the change. | PSSR reviewers must have the experience and expertise to conduct the review of the proposed changes. Must have a culture where PSSR is completed before changes are placed in service. |

### 5.9.1    Management of Change Guideline Tools

Two important administrative guidelines with respect to managing abnormal situations are the general Management of Change (MOC) guideline and the more specific Management of Organizational Change (MOOC) guideline. Failings in these two topics alone have contributed to numerous classic process safety events, as referenced in the CCPS Book *Guidelines for Management of Change for Process Safety* 2008 (CCPS 2008c). Plant process design, chemistry, equipment, controls, and personnel will inevitably be involved in changes over the lifetime of a process. Managing these changes properly can make the difference between operating a process with a lower incident likelihood versus a process with a high likelihood. Another CCPS book, *Guidelines for Risk Based Process Safety* 2007 (CCPS 2007a) has included managing changes to processes over the life of the facility as one of the nine elements in the RBPS pillar of managing risk.

Management of Change (MOC) is a well- known requirement of the United States OSHA 1910.119 Process Safety Management of Highly Hazardous Materials Standard (element l) (OSHA US 1910.119) and many other International Standards. Most established chemical process organizations have a proven and effective MOC work process. However, with respect to ASM ®, a concern can arise when the following types of changes occur for process control systems:

- logic changes
- software revisions
- tuning of controllers
- process control additions/deletions
- alarm set points
- interlock trip points

These changes may not be immediately obvious or visible to front-line personnel. Although the risks associated with these changes are typically low, under abnormal situations or plant upsets they could affect the control panel operator's normal response to an upset, resulting in a more serious consequence.

The process control engineers and technicians may consider their work as normal "replacement in kind" and therefore not requiring a formal review following an established MOC work process. This can be a significant mistake for two reasons:

1) The plant control panel operators may be blind to the change, therefore only becoming aware during an abnormal situation and

2) The effect of the change could have contributed to a serious consequence that might have been prevented if the change had been reviewed with plant control board operators.

New hires require special consideration, since although they must undergo MOC training as part of their induction and may report to a mentor, they may still not appreciate the significance of a change to a system or procedure.

Example Incident 5.5 illustrates how failure to conduct an MOC involving level control and control panel instrumentation resulted in a major petroleum spill and fire.

---

**Example Incident 5.5 – Caribbean Petroleum Tank Farm Explosion and Fire**

A massive fire and explosion occurred in 2009 at the Caribbean Petroleum, CAPECO, terminal near San Juan, Puerto Rico. Company workers were unloading gasoline from an ocean tanker into the facility storage tanks. Normally the transfer would be into a single tank that had sufficient capacity to hold the entire volume of the tanker. However, on this occasion, no large storage tank was available, so a decision was made to distribute the contents across four smaller tanks. The level in the storage tanks was protected by a single layer of measurement- a float and tape mechanical measure gauge. The level gauge sent an electronic level measurement to the control room and displayed a manual reading at the tank.

On that day, the electronic measurement function of the gauge was inoperable, so the unloading process was monitored by plant operators checking the level in the field. Meanwhile, the control room operators were operating blind to the level in the tank. An estimate was made as to when a tank would be near full capacity and require switching to another tank. When the field operator returned to the tank to check the level and make the switch, the smell of gasoline and a low-lying fog were observed. The tank had begun overflowing, draining out through an open tank dike valve and into the waste treatment plant. Electric motors in the waste treatment plant then ignited the gasoline, resulting in a large fire and explosion. No injuries occurred but 17 storage tanks were destroyed, and local residents had to be evacuated.

Had the field operator been present when the overflow began, the consequences could have been minimized. However, had the field operator been present, the operator might have been overcome by the gasoline fumes and required emergency medical care.

---

**Example Incident 5.5 – Caribbean Petroleum Refining Tank Farm Explosion and Fire – (*cont.*)**

Lessons learned in relation to abnormal situation management:

- Management of Change: No MOC was conducted to manage the loss of a critical safety barrier when the level gauge on the tank stopped functioning. As a result, they did not consider the aspect of human performance issues in managing the level in the tanks.

- Process control monitoring: The lack of the level measurement in the control room placed the control room operators in a position of operating blindly.

- Abnormal situation management recognition: Conducting the gasoline transfer under these conditions was not recognized as an abnormal practice. This resulted in an underestimation of the potential consequences of an overflow.

---

### 5.9.2    Management of Organizational Change

For an organization to perform consistently at a high level, the roles within the organization must be well defined and communicated. For example, the responsibilities of the control panel operator versus the field operator must be distinct, documented, communicated, and validated. Their roles must not conflict, but rather complement one another.  Once the organization has been well defined and the roles are clearly established, all changes to the organization should be carefully considered and reviewed. CCPS has published a helpful resource entitled *Guidelines for Managing Process Safety Risks During Organizational Change*, (CCPS 2013). The book addresses many aspects of organization changes, from staffing to hierarchy changes. It also includes many examples that illustrate where an organizational change was one of the contributing factors that led to an incident.

Additionally, the CCPS book provides several checklists and activity mapping forms that can be applied when organizational changes are being considered. These checklists and forms are tools that can be appropriately applied when reviewing organization changes.

### 5.9.3   Pre-Startup Safety Review

No matter the type of MOC, a Pre-Startup Safety Review (PSSR) should be performed. The purpose of the PSSR is to ensure that the MOC was fully reviewed, the change was communicated to affected personnel, training has been conducted as needed, and documentation and records of the changes have been updated.

In summary, a variety of tools and methods should be employed as discussed in this chapter to help evaluate abnormal situations and manage their potential consequences. Incorporating these tools into the normal management practices of a facility is therefore recommended.

Chapter 6 will examine and provide guidance on how to measure and continuously improve the management system for abnormal situations at a facility.

# 6 CONTINUOUS IMPROVEMENT FOR MANAGING ABNORMAL SITUATIONS

## 6.1   GENERAL

Managing abnormal situations should be an integral part of a facility's safety management system. Properly addressing abnormal situations requires a focus on many, if not all, of the features of a typical management system, such as:

- Scope, purpose, and anticipated goal / work product
- Roles and responsibilities
- Personnel competencies and behaviors
- Procedures for tasks, practices, and methodologies
- Necessary resources and tools
- Records, reports, and other documentation
- Document management and updating, if necessary
- Communication
- Performance monitoring and measurement
- Investigation of abnormal situations, especially repeat events
- Action tracking and resolution
- Management review and continuous improvement

Many of these features have been discussed in detail in earlier chapters. Nonetheless, successful management of abnormal situations focuses primarily on four elements of RBPS: Measurement and Metrics; Incident Investigation; Auditing; and Management Review and Continuous Improvement. The relevance of these four elements to improving the management of abnormal situations is discussed in this chapter.

## 6.2    LANDSCAPE OF AVAILABLE METRICS FOR IMPROVEMENT

Process safety performance is an important aspect of reviewing and improving how a facility manages abnormal situations. Potentially numerous leading and lagging metrics can be applied to measure factors that can cause, influence, contribute to, mitigate, and prevent abnormal situations. Since it is not practical for plants to measure everything, they should identify and focus on the specific items or areas that represent the weakest barriers and/or highest risks that could lead to an abnormal situation and a process safety incident. Methodologies for selecting appropriate metrics are discussed in *Guidelines for Risk Based Process Safety; Guidelines for Process Safety Metrics; Guidelines for Integrating Management Systems and Metrics to Improve Process Safety Performance* (CCPS 2007a, 2009, 2016c); *Cheddar or Swiss?  How Strong are your Barriers?* (Broadribb/GCPS 2009); *Process Safety Leading and Lagging Metrics ... You Don't Improve What You Don't Measure* (CCPS 2018d); *Process Safety Performance Indicators for the Refining and Petrochemical Industries* (API 2016); and *Process Safety - Recommended Practice on Key Performance Indicators* (IOGP 2018).

Many relevant examples of leading and lagging indicators are related to asset integrity and reliability. Safety-critical devices are essential for the mitigation and prevention of abnormal situations that could result in process safety incidents and must be reliable so that they work on demand. Specific examples of asset integrity metrics that facilities should consider monitoring include:

- Overdue tasks for inspection, testing, and preventive maintenance (ITPM)

- Failure of safety-critical elements/equipment, such as instrument trips and alarms, pressure safety valves (PSVs), and safety instrumented systems, to work on demand that could give rise to abnormal situations

- Number of loss of primary containment (LOPC) spills and releases due to piping and equipment failures, or maintenance practices.

Other examples of metrics (both leading and lagging) that facilities should consider include:

- Overdue reviews of operating procedures, safe work practices, and maintenance practices that could lead to abnormal situations
- Number of times upper and lower operating limits are exceeded
- Overdue training, especially for operators, on troubleshooting and managing abnormal situations
- Number of abnormal situations that occurred that were not already covered in HIRA studies
- Duration of inhibited or bypassed safety-critical elements/
- equipment, instrumentation, and alarms that are essential for warning and managing abnormal situations
- Number of outstanding recommendations from management of change, auditing, and incident investigations that are relevant to reducing abnormal situations
- Incident and near-miss rates, especially high potential near-misses, arising from abnormal situations
- Number of repeat incidents associated with abnormal situations
- Number of risk evaluations completed and number of risks requiring action identified
- Number of risks requiring action that have been mitigated versus the total number identified
- Percent of control panel operators trained on recognizing abnormal process situations
- Percent of control panel operators trained on the alarm system and individual key alarms
- Number of alarms suppressed
- Number of alarms that are in constant alarm (stale alarms)
- Number of gaps in information communicated via shift handover, from checklist-based audits
- MOCs: percent of overdue action items
- MOOCs: percent of overdue actions on organizational changes
- Number of open action items from Pre-Startup Safety Reviews (PSSRs)

Many other metrics could be included to measure the effectiveness of the facility's management systems in identifying and preventing process conditions that could lead to an abnormal situation and a process safety incident. It is recommended that the metrics be indicators that are specific to a process unit. The metrics should specifically reflect how the unit is managing its operation safely, with an emphasis on those metrics that can help predict or prevent an abnormal situation from occurring or escalating to an event. The availability of data for potential use in metrics has increased significantly in recent years, as discussed in 5.3.3. However, it is important to ensure that the metrics are accurate and relevant, to prevent "metric overload". The use of a "dashboard" to provide a high-level management summary of metrics data is encouraged.

## 6.3    ABNORMAL SITUATIONS AND INCIDENT INVESTIGATIONS

Incident investigation is a way of learning from incidents to identify management system issues and weaknesses that can be corrected, in order to improve the overall effectiveness of the management system. It is particularly important to investigate *high-potential* near-misses that could lead to fatalities, substantial property damage and/or environmental damage, under different circumstances. While it may not be practical to investigate *every* abnormal situation in depth, abnormal situations should be investigated in order to recommend actions to prevent, or at least minimize, their occurrence in the future.

A near-miss event can result in a serious process safety incident under slightly different conditions if the underlying cause is not determined and action taken to prevent it from happening again. Failure of safety-critical equipment/elements such as pressure safety valves and safety instrumented systems to work on demand, for example, can rapidly exacerbate an already serious abnormal situation if operations personnel are slow to intervene.

Like any incident investigation, the depth of analysis should be commensurate with the actual and potential severity of the abnormal situation. Several sources of guidance are available for determining and conducting the depth of investigation: *Guidelines for Risk Based Process Safety* (CCPS 2007a); *Guidelines for Investigating Process Safety Incidents* (CCPS 2019); *Pressure Equipment Integrity Incident Investigation* (API 2014);

and *Guide for Fire and Explosion Investigations* (NFPA 2021). The metrics guidance on classifying incidents can also be used to determine severity (CCPS 2018d, API 2016). Whichever approach is taken, the goal should be to identify the system weaknesses and conditions causing the abnormal situation, and then improve the system to prevent the conditions that could otherwise lead to future abnormal situations.

## 6.4  AUDITING

Auditing process safety is another way of identifying and correcting management system issues and weaknesses ***before*** they result in abnormal situations and, potentially, serious incidents. Both internal/self-audits and external audits should be conducted. Most, if not all, of the elements of RBPS (CCPS 2007a) should be audited periodically, and actions taken to improve the effectiveness of the management system, with a particular emphasis on abnormal situations.

For example, an audit may find that hazard identification and risk analysis (HIRA) studies did not address the circumstances of past occurrences of abnormal situations. Such a key finding should spur leadership to look closely at how HIRA studies are conducted, in terms of scope, methodology, and team competency / experience. Future HIRA studies, especially Hazard and Operability (HAZOP) Studies, should specifically address the management of abnormal situations.

An audit might identify that the safety systems in a plant are not capable of performing adequately (if at all) during abnormal situations. Field inspections of equipment may identify several possible issues such as isolation valves incorrectly closed or valves that are open but should be closed, missing valve manual actuators, leaking air lines to solenoids, or open-ended lines. Auditors can review the alarm summary with control room operators. Any standing/stale alarms can be discussed and measured, providing an opportunity to review the operator training and their level of understanding of the control/alarm systems and required responses. These kinds of issues can then be addressed, and the data used as a leading indicator in the metrics. Example Incident 6.1 and Example Incident 6.2 are good examples of incidents in which audits have identified deficiencies in safety systems.

**Example Incident 6.1 – Fire Protection System Found Disabled**

During a field review of safety systems in one unit, the auditor noted that the pumps were provided with a sprinkler system that was intended to provide a rapid response to a local fire. Curious as to the source of this water, the auditor followed the supply piping back to a manifold. This manifold had several valved lines tying into it, so that it was not immediately clear which valve would lead to which user. Furthermore, a painter had left a thick layer of paint on all the valves, so that a significant effort would be needed to operate the valves in an emergency. Beyond that, the auditor traced the source of the manifold's water to yet another manifold, where those valves were blocked in. The condition and position of the two sets of valves were in themselves not an emergency situation; however, they were important to the function of a mitigation system to address an emergency. In the event of an actual fire, the response would have been delayed and ineffective.

Lessons learned in relation to abnormal situation management:

- Understanding abnormal situations: A lack of knowledge about the positioning of the block valves, their readiness, and functionality. The valving was abnormal, yet the consequences were not considered.

- Process Monitoring: Continually monitor the readiness of safety systems or include confirmation that a required periodic checklist has been performed.

- Car Seal Management: The utilization of a car seal open and car seal closed system to ensure valves are maintained in their preferred position is a positive way to minimize valving errors.

---

### Example Incident 6.2 – The Dike That Wasn't

During a field review of safety systems in a tank farm, the auditor noted that the diked area around one tank had a drain valve that was open. He was told that the valve was provided to allow rainwater to be drained from the diked area following heavy rains. If the valve is not reclosed after draining the rainwater, the dike will not be able to contain a large spill from the tank – which, given the topography of the facility, would have resulted in a stream of fuel oil running downhill into the nearby process areas, creating both safety and environmental problems.

Lessons learned in relation to abnormal situation management:

- Understanding abnormal situations:    Diking of tanks for containment with installed dike drains is a standard design. The potential consequences of leaving the drain valve open should be explained and understood.

- Procedures:  A written policy or checklist procedure should be in place for managing drain valves in dikes.

---

Previous chapters have highlighted the importance of other RBPS elements with respect to abnormal situations. The intent is not to duplicate that text, but auditing should prioritize:

- Operating procedures – readily accessible troubleshooting guidance to correct abnormal situations

- Training & performance assurance – competency of workforce to recognize and intervene to correct abnormal situations

- Asset integrity & reliability – absence of overdue inspection, testing and preventive maintenance (ITPM) tasks on safety-critical equipment/elements to enhance reliability to work on demand

- Management of change – projects should identify conditions that could lead to abnormal situations, as a result of change

- Conduct of operations – behaviors, alertness, and diligence of workforce to recognize and intervene to correct abnormal situations.

Process safety engineers should consider developing specific questions on these issues for HIRA team members to use to address abnormal situations. Other RBPS elements discussed in previous chapters may also be beneficially audited with a focus on abnormal situations.

## 6.5   MANAGEMENT REVIEW AND CONTINUOUS IMPROVEMENT

Management review is the routine evaluation of whether management systems are performing as intended (CCPS 2007a). This ongoing due diligence review by leadership fills the gap between daily work activities and periodic auditing. Weaknesses and inefficiencies in a management system may not be immediately obvious, but the management review process provides regular checks so they can be identified and corrected before they are revealed by an audit or an incident.

A study by the ASM® Consortium of the root causes of 42 incidents found that the top ten causes accounted for 71% of all the operations practice failures (Bullemer & Laberge 2009). Over 10% of these top ten causes were associated with an ineffective continuous improvement program.

The management review should be led by the facility manager and involve key subject matter experts, including a senior operator and maintenance technician. The team should focus on a few RBPS elements (typically up to three) at each meeting and then evaluate and discuss records and observations pertaining to management system weaknesses associated with those RBPS elements. Management should identify weaknesses, and make recommendations for improvement and then capture them in an action plan that specifies responsible parties and completion dates. The next management review should then focus on action plan progress to drive continuous improvement, before reviewing the performance of other RBPS elements.

Management review can also be applied to improve how abnormal situations are addressed by the facility. However, the review should be part of a larger review process that addresses *any* weakness in RBPS elements, including those that are related to abnormal situations. The depth and frequency of each management review should be governed by past incidents and abnormal situations, in addition to results obtained through auditing, metrics, and previous reviews, and management's view of perceived risk posed by abnormal situations.

## 6.6    SUMMARY

This chapter has highlighted the importance of integrating the management of abnormal situations within a facility's existing safety management system, as many of the RBPS elements can be applied to reduce abnormal situations. The aim should be continuous improvement in how a facility addresses abnormal situations. To this end, four elements of RBPS (metrics, incident investigation, auditing, management review & continuous improvement) are the primary elements that can deliver continuous improvement.

Chapter 7 provides an in-depth review of three case studies detailing serious events that could have been avoided by the implementation of a continuous improvement process for managing abnormal situations.

# 7 CASE STUDIES/LESSONS LEARNED

This book contains a series of embedded example incidents that illustrate some of the key issues associated with managing abnormal situations. Using example incidents and case studies in discussions and formal training sessions can be highly beneficial in helping staff to understand the underlying causes and learnings arising from these types of events. Questions to ask staff include:

- How would you respond?

- How would you ensure that people are out of harm's way?

- How would you decide when to shut down operations?

- What do you think we could do differently to avoid a situation like this from occurring here?


Case studies are available from numerous sources, including newsletters, incident reports, and various databases as follows:

- *The Process Safety Beacon*, produced by CCPS (CCPS website)

- *Safety Digest,* US Chemical Safety and Hazard Investigation Board (CSB 2021 news website)

- *Loss Prevention Bulletin*, produced by the IChemE in the UK (IChemE UK)

- *Safety Lore*, produced by the IChemE Safety Centre in the UK (IChemE UK)

- *Learning Sheet*, produced by the European Process Safety Centre (EPSC)

- *The ICI Safety Newsletters*, mainly issued by Trevor Kletz (Kletz T)

- Health and Safety Executive UK (HSE Case Studies) (HSE UK)

- Chemical Safety Board - reports and videos on major incidents (CSB website)

- European Commission Major Accident Reporting System—a searchable database of incidents in the EU (eMARS database)

- The Bureau for Analysis of Industrial Risks and Pollutions (BARPI, 2018)—Analysis, Research and Information on Accidents (ARIA) database—a searchable database of incidents and other reference material) (BARPI database)

- CCPS Process Safety Incident Database [PSID] (CCPS 2018-2) (CCPS/PSID Database)

This chapter includes a detailed review of three case studies that are particularly relevant to managing abnormal situations. One is from the aviation industry and two are from the process industry.

## 7.1    CASE STUDY 7.1 – AIR FRANCE, 2009

### 7.1.1    Background

Aviation incidents, including aircraft crashes, have been a byproduct of the industry since the Wright brothers first flew a winged machine. Like process industry incidents, the earlier aviation incidents were often associated with the design of the equipment and the skills and knowledge of the operator/pilot. The aviation industry subsequently developed a very effective investigation procedure where key information is gathered and scientifically assessed, and reports are produced in stages as more information is obtained and analyzed. This results in rapid learning across the industry even before the final report is released, which is particularly beneficial if there are common flaws discovered in design, systems, procedures and/or human factors that could prevent similar incidents from occurring in the near future. Over time, lessons have been learned, the refined design of the aircraft and associated equipment has led to higher levels of reliability, and the safety of the industry continues to improve, similar to what has occurred in the process industries. In the US, all aircraft carrying 20 or more people must be fitted with a flight data recorder/ cockpit voice recorder or "black box". Similarly, the process industries now have access to much more historic data with a higher resolution, such that the information available is often on a par with that of the aviation industry.

The aviation industry began using flight simulators as early as the 1920s, and the first digital flight simulators were put in service in the 1960s. In the process industry, process simulators are becoming increasingly available for training purposes. The aviation industry has been highly regulated for many years, although it has recently started to introduce formal safety management systems (SMSs) that provide a top-down, organization-wide approach to managing safety. An "Advisory Circular" was issued to the US aviation industry in January 2015 (USDoT Advisory Circular) requiring service providers to develop SMSs and was followed by 14-CFR 119.8 Safety Management Systems, which requires an SMS to be in place by March 19, 2018 (ECFR Title 14, *Aeronautics and Space*). In Europe, the Civil Aviation Authority provided CAA CAP 795: *Safety Management Systems (SMS) guidance for organisations* in 2015 (CAA 2015).

Formal safety management systems have been in place in the process industries for many more years than in the aviation sector. For designated highly hazardous chemicals in the US, this was first mandated by the Process Safety Management (PSM) regulations 29CFR 1910.119 (OSHA PSM) in 1992. In 1984, the UK implemented the CIMAH (Control of Industrial Major Accident Hazards) Regulations; and in Europe, the 'Seveso Directive', on the control of major accident hazards involving dangerous substances, was originally published in 1996 and became law in 1999. In the UK, the Seveso Directive replaced CIMAH and the UK adopted it as COMAH (Control of Major Accident Hazards) in 1999. The CCPS produced their text *Guidelines for Risk Based Process Safety* in 2007 and the Energy Institute produced a *High Level Framework for Process Safety Management* in 2010 (CCPS 2007, EI 2010).

Despite the difference in the timing of implementing formal systems for safety management, many of the issues and features of the modern-day cockpit can involve challenges similar to those in process industry control rooms when it comes to addressing abnormal situations. These include, but are not limited to:

- Information and alarm overload.

- Increased reliance on automation.

- Less opportunity to practice reacting to abnormal situations.

- The "startle effect", when an automated system suddenly cannot control the process, and the operator has to take rapid action.

In both industries, automation is also gradually replacing many routine tasks, which can provide many benefits as well as risks when these systems fail.

Major aviation incidents that are associated with failures of such automated systems include:

- October 19, 2018, Lion Air flight LNI 043, Boeing 737-8 (MAX) crashed into the Java Sea some 12 minutes after take-off from Jakarta (Republic of Indonesia, 2018 Preliminary Report) leading to 189 fatalities. An automated system called "Maneuvering Characteristics Augmentation System" (MCAS) that was supposed to counter a "nose-up" tendency under certain conditions operated erroneously and repeatedly pointed the nose of the aircraft down. The MCAS system relied on angle-of-attack (AOA) data from a single AOS sensor that was not functioning correctly. The design of the system was flawed, and the pilots were unable to understand and manage the problem.

- March 10, 2019, Ethiopian Airlines flight 302, Boeing 737-8 (MAX) crashed into the ground some 16 minutes after takeoff from Addis Ababa, Ethiopia (Ethiopian Ministry of Transport 2019 Preliminary Report) leading to 157 fatalities. The erroneous operation of the MCAS system was a key factor again, although on this occasion the Pilot followed a procedure and switched off the MCAS/ stabilizer trim system. However, by the time he had done this, the aircraft was in a "mistrim" situation, and the pilots could not physically turn the trim wheels to correct the situation. Finally, they switched the trim system back on, likely attempting to use the automatic trim, but the MCAS cut in again and the plane nose-dived into the ground.

- January 9, 2021, Sriwijaya Air flight SJ 182, Boeing 737-500 crashed into the Java Sea just 4½ minutes after take-off from Jakarta (KNKT Preliminary Report, 2021) leading to 62 fatalities. The Flight Data Recorder showed an anomaly that indicated a gradual reduction in engine thrust from the left engine that was under control of the auto-throttle. The right thrust level remained unchanged. A disengagement of the autopilot occurred at 10,900 ft, which would have been compensating for the significant difference in thrust, followed by the sudden roll of the plane to the left to more than 45 degrees of bank before it crashed into the Java Sea. Prior to this flight, there had been at least two reported problems with the auto throttle. More details are awaited as the cockpit voice recorder was only just recovered (31 March 2021) at the time of writing.

In contrast to these incidents, despite suffering dual engine failure when it struck a flock of birds after taking off from New York City's La Guardia Airport in 2009, US Airways Flight 1549 landed successfully on the Hudson River. This was a situation where Captain Chesley Sullenberger used his basic piloting skills and significant experience, rather than relying on instruments.

### 7.1.2   Incident Overview – Air France AF 447

This case study concerns an incident that occurred on June 1, 2009, involving Air France flight AF 447. This Airbus A330-203 crashed into the Atlantic Ocean about 3 hours 45 minutes after take-off from Rio de Janeiro Galeão Airport bound for Paris Charles de Gaulle Airport, leading to 228 fatalities. The aircraft was at a cruise altitude of about 35,000 feet when it encountered turbulence and a high-level cloud mass. The autopilot and autothrust "disconnected" and the pilots were unable to control the aircraft, which crashed into the ocean about 4½ minutes later.

The French Bureau of Enquiry and Analysis for Civil Aviation Safety (BEA) investigated the accident and released the final report in July 2012, three years after the crash (BEA Final Report 2012).

The report identified the blockage of pitot tubes, which were responsible for speed measurement, as the first of a series of events that led to the accident. Three sets of pitot tubes on the aircraft are used to determine key flight parameters including speed and altitude. Ice blockage of the pitot tubes caused inconsistencies in the aircraft speed measurement, which resulted in disengagement of the autopilot and led the airplane to a stall position. The crew failed to recover the aircraft from the stall position.

### 7.1.3   Speed Measurement on A330 Aircraft

The airspeed on most aircraft, including the A330, is deduced using two sets of pressure data from outside the aircraft. The first is taken from a static pressure sensor, oriented flush along the aircraft surface and the second is from a dynamic sensor comprising a forward-facing tube, called a pitot tube. The difference between these two pressures, corrected for air density/height, is used by the flight computer to calculate the velocity of the aircraft relative to the air.

The A330 features three sets of pitot tubes ("probes" – see Figure 7.1) (BEA 2012) and six static pressure sensors. The probes are fitted with drain holes to prevent water from accumulating and are heated to prevent icing. In December 2001, Air France had its first delivery of the A330 aircraft, which were all originally fitted with pitot probes manufactured by Thales, model number C16195AA. The A330 on flight AF447, registration F-GZCP, first went into service in April 2005.

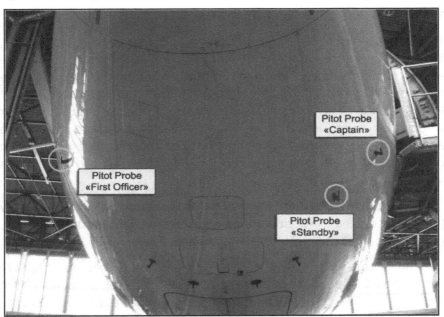

**Figure 7.1. The Three Pitot Tubes on the A330 Aircraft**

## 7.1.4 A330 Flight Control Systems

The signals from each of the three sets of probes feed into three independent computer systems referred to as "Captain", "First Officer", and "Standby" that supply three Air Data Inertial Reference Units (ADIRU) as listed in Table 7.1.

### Table 7.1 Flight Control Computers

| Pitot tube/ static tube | System name | Computer |
|---|---|---|
| 1 | Captain | ADIRU 1 |
| 2 | First Officer / Copilot | ADIRU 2 |
| 3 | Standby | ADIRU 3 and ISIS* |

* ISIS: Integrated Standby Instrument System

The aircraft uses an Electronic Flight Control System (EFCS) with "fly-by-wire" flight controls controlled by pilot input from "sidesticks", one for the pilot who is flying the aircraft (PF) and one for the pilot not flying (PNF), or copilot. The sidestick movements are transmitted in the form of electrical signals to flight control computers. This aircraft has three flight control primary computers, called FCPC or PRIM, and two flight control secondary computers, called FCSC or SEC. One of the computer system's tasks is to calculate the position of the various control surfaces as a function of the pilot's orders. The system is designed to prevent excessive maneuvers from the controls, or exceedance of the safe flight envelope. Flight control surfaces are all hydraulically activated and controlled electrically. There also is a mechanical backup for the longitudinal control pitch via pitch trim wheels and for lateral control via rudder pedals.

Under normal flying conditions, the EFCS processes the inputs from the sensors, pilots, and control surfaces using "Normal Law". This includes feedback from the aircraft response back to the control computers that command the position of the control surfaces. The pilot input via the sidestick provides signals to change pitch or direction, although the control system automatically trims the aircraft. Normal Law mode includes protection for excessive pitch, angle of attack, bank angle, and speed.

If at least two sensors or other parts of the control system fail, the system automatically reverts to "Alternate Law 1", "Alternate Law 2" or "Direct Law". These modes provide, respectively, progressively lower levels of control and protection for the aircraft, requiring more input from the pilots to control the aircraft. In both Alternate and Direct Law, the system that automatically protects against high angle of attack (that could, for example, lead to a stall), is not available, although a stall warning is still triggered above a certain threshold.

A voting system used to calculate the airspeed normally averages the inputs from the three pitot tubes. If one of the inputs deviates too greatly from the other two, however, it is rejected and the average of the two others are used. Similar systems are used on safety-critical instruments in the process industries to improve reliability. If the difference is too great between these two then this triggers the control system to switch to Alternate Law. The switch to Alternate Law also takes place if the voted value for airspeed drops by more than 30 knots over 1 second.

The static pressure measured by the probes is also used to calculate the altitude of the aircraft. However, a correction must be applied by the computer since the measured static pressure overestimates the actual static pressure, and this correction is related to airspeed. Thus, a problem with the probes cannot only lead to inaccurate airspeed but also a deviation in the calculated altitude. A sharp drop in apparent airspeed due to pitot icing, for example, would lead to a drop in the indicated altitude. In the process industries, this might be the equivalent of a situation where critical safety systems do not have independent protection layers (IPLs).

The PF and PNF have duplicate display units that comprise a Primary Flight Display (PFD) and a Navigation Display (ND). Each PFD includes displays for attitude, airspeed, altitude, vertical speed, and heading. However, each PFD uses Computed Air Speed (CAS) based on input data that originates from a different computer (ADIRU 1 or 2) and therefore a different pitot tube. The flight data recorder (FDR) is only able to save the data from two of the systems: the display on the left side PFD, which in this case was the one for the PNF, which could be from ADIRU 1 or 2, and the ADIRU 3/ISIS (standby system).

This information only provides an overview of some of the control systems that are relevant to this discussion. There are many other features and components of the flight control systems available in the report (French BEA Final Report).

### 7.1.5 Airbus Pitot Tube History

There had been a history of problems with icing of pitot tubes on some of the Airbus aircraft. In September 2007, Airbus issued a service bulletin SB A330-34-3206 (Rev. n°00) (Airfrance/Airbus Bulletin) which recommended, on an optional basis, the replacement of the THALES C16195-AA pitot tubes fitted onto all the A320/A330/A340s aircraft with a new THALES P/N C16195-BA model. This was said to improve the performance of the probe by limiting water ingress during heavy rain and reducing the risk of probe icing. At that time, only the A320s had experienced problems with the pitot tubes, so Air France's technical teams decided to modify only the A320 fleet. They planned to replace the probes on the A330/A340s only when a failure occurred, as, at the time, those aircraft had experienced no incidents involving inconsistencies in speed data. In November 2008, Airbus changed its position on the probes and stated that the "BA" model did not improve the situation with icing.

By April 2009, a total of *eight* incidents of probe icing on the A340s and one on an A330 had been reported. Additionally, the probe manufacturer, Thales, had conducted its own tests that indicated the "BA" model was indeed an improvement, so Air France decided to replace all "AA" probes with the "BA" immediately. The first batch of "BA" probes arrived at Air France on 26 May 2009 and their first aircraft was modified on 30 May 2009, just two days before AF 447 crashed, which had not yet been fitted with the new "BA" probes.

### 7.1.6 The Incident - Air France AF 447

At 22:29 on 31 May 2009, Airbus A330-203, registered F-GZCP, operating as Air France flight AF-447 took off from the Rio de Janeiro Galeão Airport bound for the Paris Charles de Gaulle Airport. Three flight crew (Captain, First and Second Co-Pilots), nine cabin crew and 216 passengers were on board. Initially, one of the Co-pilots was flying the aircraft and the Captain was the "Pilot not Flying".

By midnight, the aircraft was flying on autopilot and autothrust at a cruising altitude of 35,000 ft.

## DANGER AHEAD! STORMY WEATHER!

At about 01:35 on 1 June 2009, the co-pilot observed a large cloud mass ahead, also confirmed by the Captain. The crew observed that the high air temperature meant that they would be unable to climb to 37,000 ft. By this time, they were off the northeast coast of Natal, Brazil, and in an area with a climatology zone known as the inter-tropical convergence zone (ITCZ). This involves a mixing of air masses (trade winds) from the northern and southern hemispheres that can produce a band of rain showers and thunderstorms near the equator.

At 01:45, the aircraft entered a turbulent zone and the co-pilot commented that they were entering the cloud layer. The turbulence stopped at 0152 and the Captain woke the second co-pilot and stated that he was to take over.

At about 02:00 there was a handover between the two co-pilots, attended by the Captain, who then left the cockpit at 02:01:58 for a break. At 02:08:41, the pilots both noted the smell of ozone in the cockpit and shortly afterwards an increase in temperature. At 02:09:46, a background noise in the cockpit was heard and subsequently identified as the sound of ice crystals impacting the fuselage.

## LOSING CONTROL: THE AUTOPILOT DISCONNECTS

At 02:10:05, first the autopilot and then the auto-thrust controls suddenly disconnected and from this time onward, the flight control law switched to "Alternate". The pilot flying the aircraft, in the right seat (PF) said, "I have the controls". The stall warning triggered and stopped several times until the end of the flight. The co-pilot who was not flying the aircraft (PNF) stated, "We've lost the speeds" and then "alternate law protections".

The automatic disengagement of the autopilot appeared to coincide with a lateral gust causing the aircraft to roll to the right. The PF made several "rapid and high amplitude roll control inputs" and nose-up inputs via the sidestick, increasing the pitch of the aircraft from about 1° to 11° in ten seconds from 02:10:07. The first of two stall warnings sounded at 02:10:10; a second one was triggered 41 seconds later. The aircraft altitude was rising at a very high rate of up to 7,000 ft/minute, reaching a maximum altitude of almost 38,000 ft at about 02:11:10. Figure 7.2

shows the PF (First Officer "F/O") input from the stick in blue and the actual pitch of the aircraft in brown (BEA 2012).

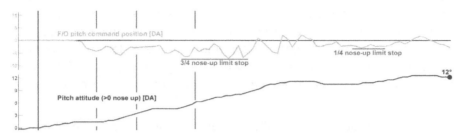

**Figure 7.2  Aircraft Pitch Commands and Pitch Attitude from 02:10:05 to 02:10:26**

At 02:11:33, the PF stated, *"I don't have control of the airplane any more now."*

At about 02:11:42, the Captain re-entered the cockpit, 1½ minutes after the autopilot disconnect. By this time, the aircraft was in a rapid descent and at 02:11:43, the PNF stated, *"What's happening? I don't know I don't know what's happening"*.

Two seconds later, the PF stated, *"We're losing control of the aeroplane there"*.

After 2:11:45, the stall warning triggered another ten times, two of which coincided with a pitch-up input by the PF. At about 02:12:00, as the aircraft dropped through 31,500 feet, the angle of attack was around 40 degrees (nose up) and the altitude was dropping at a rate of 10,000 ft/min. In the period 02:12:00 to 02:14:07, the aircraft dropped to 4,000 feet, with the pilots pitching the aircraft up and attempting to regain control.

From 2:14:17, the Ground Proximity Warning System (GPWS) "sink rate" and then "pull up" warnings sounded, and the recordings stopped at 2:14:28. The final data from the flight recorder showed a descent rate of 10,912 ft/min, a ground speed of 107 knots and a pitch attitude of 16.2 degrees nose-up. No emergency message was transmitted by the crew.

## THE CAUSE OF THE AUTOPILOT DISCONNECT– ICE BLOCKAGE

The freezing and blockage of one of more pitot tubes led to a discrepancy in the airspeed from the three pitot tubes and the disconnect of the automated flight control systems. This was due to a combination of the atmospheric conditions and the design of the pitot tubes that had an evolving history of blockage due to ice. The report stated that this was a known, but misunderstood phenomenon at the time.

The expectation was that pilots would be able to recognize what had happened and take appropriate action using a standard procedure. However, this did not happen for this event, nor for several of the previous occurrences of pitot tube blockage.

The recovered flight data recorder (FDR) shows the CAS that is available to the PNF, irrespective of the source of the data that can be switched between ADIRU 1 and 2, and the CAS from the ISIS standby system. The data displayed on the PF's display is not recorded.

It can be seen from Figure 7.3 (BEA 2012) that the CAS diverged significantly at about 02:10:06, briefly came together again at 02:10:17 and then diverged from 02:10:34, before coming together again at about 02:11:07. This was a result of the pitot tubes freezing and unfreezing at different times, due to the extreme icing conditions outside the aircraft, associated with the meteorological conditions.

This trend data, obtained from the FDR, is not available to the pilots; they are provided with instantaneous readouts via their respective screens, which may have been displaying different data derived from different pitot tubes (Tube 1 or 2).

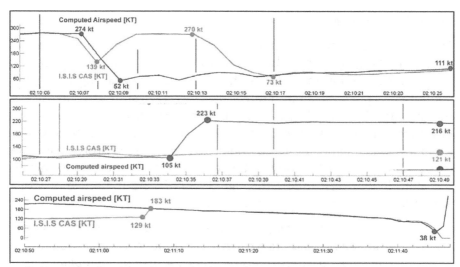

\* Note: KT or kt = knots (nautical miles per hour), 1 kt = 1.15 miles/hour, or 1.85 kilometers/hour.

**Figure 7.3  Airspeed Indication from 02:10:06 to 02:11:46\***

## MORE PROBLEMS

At 02:10:05, the pilots were suddenly responsible for controlling the aircraft, under stormy conditions but were unable to work out what was happening. The report indicated that the PF's action to pitch the nose up might have been a response to his desire to regain cruise altitude, since the instruments indicated an apparent drop of 300ft and a descent rate of 600ft/minute (actually an erroneous readout associated with the pitot tube blockage). The report also stated that the sudden disengagement of the autopilot and auto throttles coincided with a gust that increased the roll angle from 0 to 8.4 degrees in 2 seconds. The PF responded with large inputs on the control sidestick, including a nose-up input that the report described as abrupt and excessive. *The excessive amplitude of these inputs made them unsuitable and incompatible with the recommended aeroplane handling practices for high altitude flight.* Without the automated protection of the Normal Flight Control Laws, the input led to large rolls and upward pitching of the aircraft.

## THE FINAL DESCENT

The continued nose-up attitude, coupled with a reduction in actual airspeed when the aircraft was close to the edge of its safe operating window, triggered a stall. Despite the repeated stall warnings, the pilots continued to pitch the aircraft upwards. The report states that in the first minute after the autopilot disconnected, the aircraft was outside its flight envelope. Once the Captain entered the cockpit, the aircraft was in a rapid descent, and he appeared unable to diagnose the problem in time. By 02:12:00, some 18 seconds after the Captain re-entered the cockpit, the altitude was descending through 31,500 ft and the angle of attack was about 40 degrees. The report states:

> *Only an extremely purposeful crew with a good comprehension of the situation could have carried out a maneuver that would have made it possible to perhaps recover control of the aeroplane.*

The FDR stopped recording at 2:14:28, presumably when the aircraft struck the surface of the ocean.

### 7.1.7    Lessons Learned Relevant to Abnormal Situation Management

There were many lessons learned in relation to the aviation industry, most of which are also directly relevant to the process industries. The following section details are some of the key lessons that can be extracted from this case study, using some of the same categories and terms that were used in earlier parts of this book; it is suggested that the reader reviews the full report to obtain the most benefit from all of the learning and recommendations.

### *7.1.7.1    Design and HMI:*

Flight control systems are, by their nature, complex systems and require a high level of reliability. They typically have several layers of redundancy, as was the case with the airspeed measurement. Problems with the blockage of the pitot tubes due to icing under certain meteorological conditions had occurred in the past. This had led to disconnection of the automatic controls but no serious incidents. An improved design of the hardware was available and was being fitted to the fleet but had not been fitted to this particular aircraft.

The associated HMIs were also key factors including:

- The PF and PNF have displays with airspeed indicators from different pitot tubes, with the potential to cause confusion when they provide different readings.

- It is not possible for the PF or the PNF to observe the position of both sidesticks simultaneously. Thus, unless the pilots communicate clearly, it is difficult for one pilot to understand the other's control input.

- Several alarm messages indicated on the ECAM (Electronic Centralized Aircraft Monitoring) system, and the report stated that these were read out by the PNF in a "disorganized manner". Seven lines are available on the ECAM for message display and once those lines are full, a green arrow points downwards to indicate other messages of lower priority that have not been displayed. To view those messages, the pilot must clear the earlier messages, although it was not possible to determine if any of the crew cleared one or more ECAM messages during the incident. No announcement, however, to this effect was made.

- The report states: No ECAM message enabled a rapid diagnosis of the situation to be made initiating the appropriate procedure.

- Information on the angle of attack is not directly available to the pilots.

- The sidesticks have no artificial feel but they do have a spring centering device when the stick is released. Theoretically, both the PF and the PNF can make simultaneous inputs to the sidestick, in which case, the flight system sums the input from both sidesticks, up to a pre-set limit. Simultaneous inputs from both sidesticks would trigger an audible alarm and a light on the instrument panel, although there was no report of simultaneous inputs being made during this incident. Nevertheless, if the PF and the PNF had different understanding of the situation, it would be difficult for the PNF to know the control input to the sidestick without good verbal communication.

The loss of the Normal Flight Control laws meant that the systems that usually prevent the pilots from making control changes outside the operating envelope no longer existed. This is even more critical at high altitude. A combination of lack of feedback from the sidestick and this loss of protection made it more likely that excessive changes to the flight control surfaces would occur.

The design of the HMI did not make it easy for the crew to assimilate the information and form an accurate mental model of what was happening to the aircraft. Improvements to the HMI design including a display of angle of attack were recommended in the report.

### 7.1.7.2 Abnormal Situation Recognition:

Clearly, as soon as the automatic systems disengaged, the aircrew were aware of an abnormal situation. However, they were unable to diagnose what was happening to the aircraft within the available time, which was a matter of some two minutes. The instruments presented them with conflicting information: the altitude suddenly appeared to decrease, but so did the speed, although the measurements were inconsistent. Although the display included an artificial horizon, it did not include the angle of attack. The action to gain height is understandable although in hindsight, this was the wrong action to take and led to the stall. The stall indicators sounded repeatedly, but the PF did not put the nose of the aircraft down, except very briefly.

Similar situations can occur in the process industries when there are sudden changes in the weather conditions. For example, if instrument air systems are not kept very dry (typically to a dew point of -40 °C (-40 °F), moisture can freeze in lines or in mechanisms leading to erroneous readings and stuck valves.

### 7.1.7.3 Human Factors and Crew Resource Management:

It is typical with incidents of this nature for Human Factors to contribute to most of the causes. The startle effect was a key factor that led to the large input to the sidestick and the aircraft rapidly coming out of a safe flight envelope. The sudden increase in workload led to a degradation of the communication between the pilots. The report refers to the surprise generated by the autopilot disconnection and the loss of cognitive control of the situation. The report highlighted that the initial and refresher training provided did not adequately address this type of sudden scenario and recommended improvements in this area including reinforcement of Crew Resource Management (CRM) training and improved training simulators.

### 7.1.7.4    Procedures:

The report refers to flight procedures that should be followed in the event of unreliable airspeed indication. This provides a table of flight settings to ensure that the aircraft operates within its safe flight envelope. This procedure does not appear to have been followed by the pilots, although following analysis with reports and statements from other crews, it states:

> ... although technically adequate, details of the procedure continue to be understood to differing degrees by crews, who do not always consider their application necessary, and even sometimes consider them to be inappropriate at high altitude... Some crews mentioned the difficulty of choosing a procedure bearing in mind the situation (numerous warnings).

The development of procedures should always involve the people who are using them, to make them accurate, meaningful, and usable.

### 7.1.7.5    Communications:

During the handover before the Captain left the cockpit, he did not specify which of the two co-pilots would be his designated relief, nor did he provide any instructions for crossing the ITCZ. In particular, he did not comment on the meteorological situation which was about to be encountered during the ITCZ crossing. He also did not provide instructions concerning the tactics for crossing the ITCZ, or on the PF's decision to try to climb above the cloud mass. This may have given rise to possible issues regarding hierarchy in the cockpit between the two pilots and perhaps among the three of them after the Captain returned.

The report referred to a deterioration in the quality of the communications between the PF and the PNF, associated with the stress of the situation.

Training in simulators using suitable, realistic scenarios can provide useful experience for pilots in stressful situations.

### 7.1.7.6    Training/Knowledge & Skills:

When the autopilot disconnected and the control laws were reduced, the aircraft was stable, but the sudden introduction of control inputs rapidly brought it outside the flight envelope. Pilots are not used to flying under such conditions and the report recommends improved training for pilots that takes into account all configurations of the flight control laws and improves their basic understanding of fundamental flight mechanics.

The report recommends use of higher fidelity simulators that are more adept at reproducing this type of scenario, including provision for the element of surprise.

The report stated that flight crews did attend CRM refresher training twice per year, including simulator training and evaluation of non-technical skills (NOTECHS) in practical situations. However, the staff passed their training despite poor CRM performance.

Where processes are controlled with greater degrees of automation, it is easy to forget how to operate them without such protection, which should be reflected in the refresher training program.

### 7.1.7.7    Learning from Experience:

There had been several similar events involving freezing of pitot tubes prior to the crash of AF 447. It was clear that the previous events had led to confusing situations for the aircrew. The report studied 13 previous events involving unreliable airspeeds and the actions taken by the pilots—in none of the previous events did the crew follow the "Unreliable Airspeed" procedure.

The learning from these previous events, which could be classified as "near-misses", led to a program to replace the pitot tubes with a different design that was less prone to freezing. However, this program was not sufficiently expeditious to allow for timely replacement of the pitot tubes on the AF 447 aircraft.

### 7.1.8    Epilogue

Following the AF 447 incident, on 12 August 2009, Airbus issued a "Mandatory Service Bulletin" SB A330-34-3231 describing procedures for the replacement of the THALES probes with at least two GOODRICH PN 0851-HL probes. This was made mandatory by the European Union Aviation Safety Agency (EASA) on 7 September 2009, with a 4-month compliance timeline, via Airworthiness Directive 2009-0195 (EASA website) that states:

*Occurrences have been reported on A330/340 family aeroplanes of airspeed indication discrepancies while flying at high altitudes in inclement weather conditions. Investigation results indicate that A330/A340 aeroplanes equipped with Thales Avionics pitot probes appear to have a greater susceptibility to adverse environmental conditions than aeroplanes equipped with Goodrich pitot probes.*

*A new Thales Pitot probe P/N C16195BA has been designed which improves A320 aeroplane airspeed indication behaviour in heavy rain conditions. This same pitot probe standard has been made available as optional installation on A330/A340 aeroplanes, and although this has shown an improvement over the previous P/N C16195AA standard, it has not yet demonstrated the same level of robustness to withstand high-altitude ice crystals as the Goodrich P/N 0851HL probe. At this time, no other pitot probes are approved for installation EASA AD No.: 2009-0195 EASA Form 110 Page 2/3 on the A330/A340 family of aeroplanes.*

Other aviation agencies around the world issued a similar directive at approximately the same time the EASA directive was issued.

## 7.2 CASE STUDY 7.2 – TEXACO REFINERY, MILFORD HAVEN, WALES, JULY 1994

### 7.2.1 Background

An explosion and fire at the Texaco Milford Haven, Wales facility occurred on 24 July 1994, some eleven years before the better-known Texas City explosion in 2005. The causal factors, root causes and learning opportunities from the two incidents are unexpectedly similar, however. Outlined in Chapter 2, Example Incident 2.3, the Milford Haven incident is also included in this section as a more detailed case study, rather than the Texas City event, because the learning from it is perhaps even more relevant to the management of abnormal situations. The Milford Haven incident began with a severe thunderstorm that caused electrical disturbances and a lightning strike that led to a small fire on the crude distillation unit. A common misunderstanding is that this initial fire led to the larger explosion. However, the disruption caused by the power outages and the subsequent actions taken, coupled with many other issues including faulty instrumentation, historic changes, control system design, HMI and human factors actually led to the explosion some six hours later.

The consequences of the explosion at Milford Haven were much less severe than at Texas City, partly due to good fortune. The incident occurred on a Sunday when far fewer staff were present on site. Furthermore, some individuals had just left a blockwork building that was being used to write permits, which collapsed shortly afterwards following the explosion. Nevertheless, twenty-six site personnel suffered minor injuries. The physical repairs to the refinery cost some £48M ($68M), excluding business interruption losses. The refinery was prosecuted and fined by the UK Health and Safety Executive (HSE) under the Health and Safety at Work etc. Act 1974. This case study is based on information contained within the HSE Report into the incident (HSE UK 1997), which should be referred to if a more thorough understanding of the incident is required.

### 7.2.2    Incident Overview – Texaco Milford Haven

This case study concerns the release and ignition of flammable liquid and vapor hydrocarbons from a failed 30-inch (0.76m) flare knockout drum outlet pipe.

From the initial electrical disturbances that started at about 07:20 due to a thunderstorm, to the failure of the pipe at 13:23 and its ignition some 20 seconds later, there were a number of failed opportunities by operating staff to recognize an escalating problem on the facility and take appropriate action. However, operators were handling many issues at the facility and were focused on keeping the Fluidized Catalytic Cracker Unit (FCCU) from shutting down. At one stage, an outlet valve from a distillation column closed and failed to re-open, but the feed to the column continued until the level was well above normal. As the column level began to build, the pressure increased and relief valves opened, carrying liquid and vapors into the flare line. Problems with high liquid levels elsewhere on the unit led to further demands on the flare system, including the use of hoses to drain hydrocarbons from one vessel and into another.

Faulty and unreliable instrumentation made it difficult for operators to establish what was going on. The DCS produced hundreds of alarms, all with the same level of priority but with no overview screen that would have provided operators with an indication of discrepancies on the mass-balance.

A modification made several years before meant that an automatic pump-out facility for the flare knockout drum was not lined up for service. Previous inspection of the flare line had showed that it was thinning and it was due for replacement in the next major turnaround in the following year. However, the thickness measurements did not include the point of failure, which would have indicated that the drum was in a far worse condition than the areas that had been measured.

Venting of hydrocarbon liquid and vapors continued from several sources until the flare knockout drum was overfilled and large volumes of liquid were carried down the 30-inch piping. It failed at 13:23 at the weakened 90° elbow, the second bend downstream of the knockout drum. Up to 20 tonnes (44,000 lbs) of hydrocarbons released and ignited some 20 seconds later.

### 7.2.3   Outline Process Description of Milford Haven Refinery

The description has been taken from the more detailed HSE report (HSE UK 1997), which should be referred to if greater understanding is required. However, this extracted information should provide sufficient detail for the reader to understand the learning associated with the management of abnormal situations.

The Milford Haven refinery in South Wales includes a crude distillation unit (CDU) that processes 190,000 barrels per day and incorporates several other process units including a vacuum distillation unit (VDU), FCCU, butane isomerization unit and HF alkylation unit. The CDU separates crude oil by fractional distillation into intermediate products including naphtha, gas, kerosene, diesel, and heavier components. The heavy fractions supply the VDU that in turn feeds the FCCU, which cracks the long chain hydrocarbons into lighter fuel products including light naphtha, butanes, propanes, ethane, methane, etc. The heavier material from the FCCU includes gas oil and fuel oil.

The key process area relevant to the incident is the separation section, as seen in Figure 7.4 (HSE UK 1997) that is fed with hydrocarbons from the top of the main fractionation column, downstream of the FCCU. The overheads stream cools and passes through the primary and then the secondary overhead accumulator, F-203. Liquids collect in the accumulator and the remaining vapors are compressed in the wet gas compressor and then cooled before entering the high-pressure separator, F-310.

Liquid from the high-pressure separator is pumped to the Deethanizer column, F-302, which removes mainly C2's (ethane) at the top and the bottom product supplies the Debutanizer column F-304, which removes C3's and C4's at the top, which comprise the LPG product.

The bottom of the Debutanizer is fed to the Naphtha splitter F-305, which produced Light Cycle Naphtha (LCN) that is used to blend into gasoline product.

The flare system collects relief streams from various parts of the process and feed them to the Flare knockout drum F-319. This is designed to remove any liquids from the hydrocarbon stream, allowing gases only forward to the flare stack.

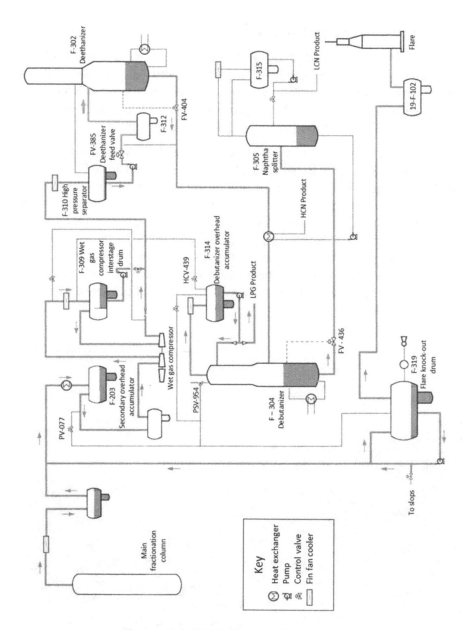

**Figure 7.4 FCCU Separation Section**

### 7.2.4   Controls and Instrumentation

The process was controlled by a Distributed Control System (DCS). A separate high integrity, independent monitoring and alarm system, called a Critical Process Controller (CPC) was used for all designated safety-critical alarms and interlocks. This included protection for the Wet Gas Compressor system such as liquid in the suction and interstage drums.

Operators controlled the FCCU unit via the DCS keyboards and six designated display units in the control room, as seen in Figure 7.5 (HSE UK 1997). It was common practice for one of the displays to show the alarm page, which could display up to about 20 of the latest high-priority alarms. The older alarms disappeared off the bottom of the page and the operator had to scroll downwards to view them. Five pages of alarms were available to the operator. On the FCCU unit, all but one of the alarms were given the same (high) level of priority. There were also several alarms displayed separately on annunciator panels, some on the control console and others on the main panel behind.

**Figure 7.5  Texaco Refinery Control Room DCS Screens**

## 7.2.5    Some Relevant History at the Refinery

Flare Knockout Drum pump-out

The flare knockout drum F-319 was originally designed to have a liquid capacity equivalent to some 20 minutes of the maximum foreseeable feed rate of liquids from the refinery relief system. It was also designed with an automatic pump-out system that would operate above a certain liquid level, pumping liquids to a slops vessel in the off-plots area at a rate of up to 221 $m^3/h$, further increasing the capacity of the system to receive liquids.

Presumably, this could accommodate an incident with a full refinery trip or fire scenario, and the time period was deemed sufficient to deal with liquids entering the flare system under these circumstances, before the unit was safely shut down. This also complied with Texaco standards at the time, which required a capacity of at least 15 minutes surge volume at maximum design liquid input and without pumping out.

In 1991, the pump-out system was modified so that the liquid would normally be pumped back into the secondary overhead accumulator F-203. This change was carried out for environmental and efficiency reasons as it allowed the liquid to be recovered directly, rather than going to slops and then being further treated. However, it meant that any liquids in the knockout drum would be looped back into the distillation section and at a rate that was limited to 7 $m^3/h$. The original system, which pumped at the much higher rate to the off-plots area, was left in place although it was manually isolated by a gate valve and required an outside operator to operate it manually. Therefore, unless this gate valve was opened manually, the pump-out time for the flare knockout drum was effectively increased from less than 30 minutes to over 15 hours.

It was thought that the original pump-out arrangement had not been used since the modification was made in 1991 and was not used at the time of the incident. The report states that operators had neither the experience nor the instructions for reconfiguring the system to allow it to pump to the slops tank.

The investigation was unable to locate any records of a risk assessment/Management of Change (MOC) process having been carried out when the modifications were made.

Flare line and its inspection

The flare pipework was subjected to thickness checks at various locations and the site was aware that corrosion had become severe over a period of months before the incident. However, they did not measure the thickness at the location that failed. They had intended to replace the flare line at the next turnaround.

Nevertheless, a flare line is not designed for a heavy liquid load; it is assumed that any liquids would be removed in the knockout drum.

### 7.2.6    The Incident

The incident began with an electrical storm that started at about 07:20 on Sunday 24 July 1994. Full details of the incident and its timeline are contained in the HSE Report (HSE UK 1997). When reviewing this incident, it may be useful to refer to Figure 7.4, which is also available as Figures 11-16 in the HSE report.

### THE BEGINNING

The electrical storm led to various power cuts between about 07:49 and 08:30, resulting in a temporary loss of feed to the FCCU followed by a reduction in feed rate from the main fractionator. This led to a loss of level in the high-pressure separator F-310 upstream of the deethanizer F-302, which fell to around 6%. To allow this level to recover, at 08:33, the operator manually reduced, very slightly, the opening of deethanizer feed control valve, FV-385, to around 36% of the indicated range, via the DCS.

It was known that control valve FV-385 was temperamental and rather than leading to a reduction in the flow to the deethanizer F-302, the flow actually fell to zero. This led to the level in the deethanizer F-302 rapidly falling, since it had lost its feed, followed by its level controller automatically closing the outlet valve FV-404, which stopped the flow into the downstream debutanizer F-304.

Consequently, by 08:39, the level in the debutanizer F-304 started to fall and its level controller also responded by closing the outlet valve FV-436. Having just been fed with lighter hydrocarbon fractions, due to the rapid emptying of the deethanizer and with no fresh, cold feed to the debutanizer, from about 08:46, its temperature and pressure began to rise rapidly.

## FILLING UP

Meanwhile, liquid continued feeding forward from the main fractionator, which started to fill up the high-pressure separator F-310. Its outlet valve FV-385 should have opened but remained closed due to a fault. By 08:52, the level in the high-pressure separator F-310 exceeded 100% of its range, leading to a build-up of pressure upstream. The pressure control valve PV-077 on the secondary overhead accumulator F-203 opened to 28% allowing it to vent to the flare system.

The pressure downstream in the debutanizer F-304 continued to increase, due to outlet valve FV-436 still being stuck closed (it is thought that it remained closed throughout). This was made worse when the reflux pumps tripped at 08:51 (Reflux provides cooling to the column). By about 08:53, the pressure reached 12.6 barg and the pressure relief valves on the debutanizer PSV-954 lifted, placing further demands on the flare system.

## REESTABLISHING FLOW

In order to reduce the level in the high-pressure separator F-310, operators then manually opened the deethanizer feed valve FV-385 from an indicated 36% to 38%. The valve was faulty and the apparent slight change led to a sudden restart of the flow with a flood of hydrocarbons into the deethanizer F-302, and a corresponding flow into the debutanizer F-304 once a level was established in F-302.

## PRESSURE RISING

Operators were apparently unaware that the outlet valve FV-436 from the debutanizer F-304 was still closed. The debutanizer level was showing 79%. However, the level gauges were all differential pressure devices, which are typically inaccurate when above full range. They can also give inaccurate readout when the composition/ density is different from the calibrating fluid. Operators noted that the pressure in the column was still rising, and at 09:30, they isolated the heat source to its reboiler. By 09:37, the pressure was getting close to the relief valve setting. They decided to vent pressure from the debutanizer via the overhead accumulator drum, F-314 through valve HCV-439 to the wet gas compressor interstage drum F-309. This was an infrequently used line, intended for vapor flow only and they opened the valve manually to 25%. Under the circumstances, with the debutanizer overhead accumulator probably overfilled, liquids flowed back with the vapors to the wet gas compressor interstage drum F-309.

This action was insufficient to reduce the pressure sufficiently and at 10:01, the debutanizer relief valves PSV-954 opened again, with the pressure peaking at 12.2 barg. This discharged more liquid into the flare header, and at 10:10, the liquid level in the flare knockout drum F-319 started to increase rapidly.

**COMPRESSOR TRIP #1**
Meanwhile, the high flow of liquid from the debutanizer overhead accumulator F-314 back to the wet gas compressor interstage drum F-309, in addition to its normal feed, led to a sudden increase in level, which overflowed the internal weir into the "dry" side of the drum F-309. This caused the compressor to trip automatically via the CPC at 10:08.

Operators were unable to empty the dry side of F-309 and carried out an improvised procedure using two steam hoses to drain the drum into the flare system.

This eventually allowed them to restart the compressor at 12:28.

**FLARE KNOCKOUT LEVEL**
From 10:10 to 10:24, the level in the knockout drum F-319 increased from an indicated 61 to 93%, equivalent to about 44 $m^3$ of pentane (exact composition unknown). At 12:56, the high-level alarm on the flare knockout drum F-319 was activated, although it does not appear that this was noticed by operators and did not reset before the incident.

## MORE VENTING, SECOND COMPRESSOR TRIP

After the wet gas compressor was restarted, the debutanizer pressure started rising again and at 12:46, the relief valves operated for the third time and continued to do so for about 40 minutes, until after the explosion.

Again, operators opened the manual vent back to the wet gas compressor interstage drum F-309 and the compressor tripped for the second time at 13:21 due to excessive liquid in the drum.

At 13:22, the pressure control valve on the secondary overhead accumulator opened to 63% to relieve the pressure that was building on the main fractionator. By this time, the flare knockout drum was filled beyond its design capacity.

At 13:23, the 30-inch diameter pipe downstream of the flare knockout drum ruptured and the escaping hydrocarbons ignited some 20 seconds later, resulting in a large explosion.

### 7.2.7 Immediate Cause

The failure was due to the overfilling of the flare header knockout drum, which led to liquid being carried downstream through pipework that was only designed to carry vapors. The piping failed at its weakest point, due to localized thinning of the piping that had not been identified by the mechanical integrity program.

### 7.2.8 Lessons Learned Relevant to Abnormal Situation Management

The majority of the underlying and root causes associated with this case study are associated with issues relating to the management of abnormal situations as discussed throughout this book

#### 7.2.8.1 Design and HMI

The design of the DCS displays allowed the operators to view the unit section-by-section but did not provide an overview screen to allow them to understand the *bigger picture* more easily. This would have led to a loss of situational awareness. Although the faulty valve on the outlet of the debutanizer, FV-436 was indicating open when it was closed, an overview screen would have made it easier to spot no flow downstream of the debutanizer.

While the DCS system allowed different levels of alarm prioritization, this feature was not used effectively. Consequently, alarms that should have had different priorities assigned were presented in the same place, leading to excessive alarms and making it difficult for operators to understand the relative importance of one alarm versus another. This is what we now refer to as "alarm management", including rationalization and prioritization as well as how the alarms are presented in the HMI.

A key action from the investigation was to identify safety-critical alarms and to present them to the operators so that they are distinguishable from less important, operational alarms. Furthermore, the report states that ultimate plant safety should not rely on an operator response and should thus require an automated response of suitable integrity, based on an appropriate hazards and risk analysis.

### 7.2.8.2    *Abnormal Situation Recognition*

At the time, the operating staff were handling several abnormal situations, including a fire on the crude unit and process upsets caused by the power supply interruptions. However, it is apparent that they did not spot the crucial problem associated with the high level in the debutanizer arising from its closed outlet valve.

Chapter 3, section 3.1.2.1, refers to the contribution instrument failures can have to abnormal situations. These types of failures can make diagnosis difficult, and in this case, problems were identified with two valves as well as anomalies with level indication.

- The valve FV-385 feeding the deethanizer appears to have closed when it was manually adjusted to 36% open by the operator.

- The outlet valve from the debutanizer, FV-436 closed automatically to maintain level but then stayed closed, even though the indicator on the DCS showed that it had opened.

- The level indicator on the debutanizer reached 79% and stayed there throughout the incident. Many of the level gauges on the plant were of the differential pressure type and these can be inaccurate, particularly if the density of the fluid changes. This issue is identical to one with the Texas City incident (Example Incidents 3.3 and 4.1) and it is especially important for operators to understand this failure mode with these types of instruments.

Despite these instrument faults, no other evidence was available to operators regarding no flow downstream of debutanizer. The instrumentation on the naphtha splitter F-305 showed that there was no level in the column and no flow leaving it. This does not appear to have been identified at the time, although the lack of an overview screen made this type of diagnosis more difficult.

### 7.2.8.3   Human Factors, Culture and Crew Resource Management

Typically, for these types of events, human factors are present throughout. The workload and stress levels increase during an abnormal situation, and it is important that operators and supervisors can maintain situational awareness and realize when the problem is getting out of control.

The focus was to deal with the various problems, that were numerous, and it does not appear that anyone took a step back to look at the situation holistically. Under these types of circumstances, there can be a tendency to focus on keeping the process going, rather than shutting it down and diagnosing the problems in more detail. The culture must be such that where there are safety concerns, shutting down a process is encouraged.

The unreliability of the instrumentation suggests a culture that tolerated such conditions. Once the assessment of alarm prioritization has been completed, it would be easier for operators to specify maintenance priorities for critical instruments. Metrics can then be used to identify problem areas.

The HSE report stated that the FCCU operating team were multi-skilled and flexible in the way they worked. When problems occurred, all personnel, including management, helped. This can be beneficial, provided individuals have a thorough understanding of what the other is doing. However, responsibility for taking key actions must be defined, based on an assessment of advice provided by the team. This kind of situation can benefit from training and exposure to the principles of

Crew Resource Management/ Non-Technical Skills training as discussed in Chapter 5, Section 5.7.3.

### 7.2.8.4   Procedures

Guidance and training should be provided on the criteria for initiating controlled shutdowns, particularly where there are safety concerns. It is not clear from the report whether operators realized the magnitude of the problems that were developing, but issues with increasing levels in a flare drum should be included as a reason for shutting down ("Stop Work Authority"). This would have to be a balanced approach as even the controlled shutting down of a process could lead to an increased load on the flare system.

The pump-out system for the flare knockout drum was modified some 3 years before the incident, which had the effect of significantly reducing the liquid pump-out rate from the drum. A manual valve was available to put the original system back in service but there were no procedures or training available to use this valve. Provision of procedures and training would normally fall under part of the Management of Changes (MOC) system.

The report recommends improvement to the mechanical integrity program for the areas of the plant where there is a high consequence in the event of failure. This would include improvement to the inspection procedure for the flare line and consideration of increased end-of-life thickness criterion to account for unexpected scenarios that could lead to increased loading.

### 7.2.8.5   Management of Change

There should have been a formal procedure associated with the modifications to the knockout drum pump-out system. The significance of this change, whereby the automated pump-out rate was significantly reduced from 221 to 7 m³/h, which also returned the liquid to the process area that originally released it, was not realized. The original design intent, with an automated system at the higher rate had become a manual system requiring an operator to open a valve for it to function. No procedure or training had been provided. The report recommended that a formal procedure to evaluate the hazards and risks of plant modifications be implemented. This would be a key element of a Management of Change (MOC) system

While not contributing significantly to the incident, the action taken by operators to drain the liquid level from the wet gas compressor interstage drum using two steam hoses coupled to the flare header should have been considered a temporary modification, requiring a formal risk assessment. Again, this would form part of a formal MOC system.

### 7.2.8.6   Communications

Due to the number and the magnitude of operating problems, a large team was in the control room providing assistance. The report states that under such circumstances, there is an even greater requirement for effective communication, to ensure that contradictory operations are avoided.

### 7.2.8.7   Training/ Knowledge & Skills

The report recommends that staff be trained to include:

- "An assessment of their knowledge and competence for their actual operational roles under high stress conditions; and

- Guidance on when to initiate controlled or emergency shutdowns and how to manage unplanned events including working effectively under the stress of an incident."

Regular training on incident and accident scenarios, particularly involving the use of simulators coupled with some "traps" of unreliable instrumentation, would be beneficial.

### 7.2.8.8   Learning from Experience

The first two recommendations on the HSE report state that:

- The safety management systems should include a means of storing, retrieving, and reviewing incident information from the history of similar plants.

- Safety management systems should have a component that monitors their own effectiveness.

As a further step, where learning from other incidents is relevant to current operations, that learning should be embedded into the way the facility operates. This could include areas such as revised operating or emergency procedures, updated inspection processes, new training procedures, and content.

### 7.2.9    Epilogue

The Milford Haven incident in 1994 preceded many of the subsequent developments involving alarm management and an understanding of control room HMI issues and human factors. Many of these matters were typical of DCS systems at the time; the system at Milford Haven was installed some four years earlier. Alarms could be added to instruments for zero cost and personnel were inclined to put alarms on every instrument, even though they may not be that important. The concept of prioritizing critical alarms, grouping alarms so that only one is displayed in the event of, for example, a process trip, identification of "bad actors" was not always considered at the time.

The Abnormal Situation Management® Consortium was established in 1994, although it grew out of some work that began in 1989 by Honeywell's Alarm Management Task Force to address alarm floods. In 1997, a report commissioned by the UK Health and Safety Executive (HSE) was issued, *"The Management of Alarm Systems"* (Bransby/HSE UK 1998). This report refers to the Milford Haven incident as well as several other incidents associated with alarm management. Further work with the HSE, the Engineering Equipment and Materials Users Association (EEMUA) and other parties led to the first (1999) publication of EEMUA 191 *"Alarm Systems - A Guide to Design, Management and Procurement"* (EEMUA 191 1999) followed by the development of the other standards and guidelines outlined in Chapter 4, Section 4.2.5.

## 7.3    CASE STUDY 7.3 – THE HICKSON AND WELCH FIRE, 1992, CASTLEFORD, UK

### 7.3.1    Background

The fire at Hickson and Welch has been included as a case study because its circumstances are slightly different from the previous two events. This incident occurred during a period of maintenance, although many of the root causes arose from abnormal situations that had developed in the months and years preceding the incident. The event occurred when process technicians were raking out residue that had accumulated in the bottom of a vessel from an open, vertical manway. The residue suddenly ignited, and a horizontal jet of flame erupted from the vessel in the direction of the nearby control building, which stood between the vessel and a four-story office block, as seen in Figure 7.6 (HSE UK 1994).

The control room was of a lightweight, timber construction and the jet flame traveled through the building, extensively damaging it, and then jetted against the outer wall of the office block, starting additional fires.

**Figure 7.6  Source of the Jet Fire with Destroyed Control and Office Block**

There were five fatalities: four in the control building and one in the office block. There were two reportable injuries (requiring hospital treatment) and fifteen employees who required treatment for minor injuries and shock.

This report is mainly based upon the report by the UK Health and Safety Executive (HSE UK 1994).

### 7.3.2    Incident Overview – Hickson and Welch fire

Problems were being encountered with the operation of a batch distillation operation that used a steam-heated vessel known as "60 Still Base," shown in Figure 7.7 (HSE 1994) and a decision was made to open it up and clean out an accumulation of sludge on 21 September 1992.

**Figure 7.7  Manway at the End of 60 Still Base - Source of the Jet Fire**

Two technicians had been working from a scaffold and raking the sludge out of the vessel for about an hour before one of them left to get a longer rake. Shortly after, the remaining technician noticed a flame inside the vessel and managed to jump from the scaffold before a horizontal jet of flame shot out of the manway. The jet lasted for less than a minute before it subsided but was sufficiently strong to project the manway cover, which was on the scaffold, towards the control building. The jet destroyed the scaffold, control building and set fire to the main office building behind.

The incident was investigated by the UK HSE and Hickson and Welch Limited was prosecuted and fined under the Health and Safety at Work etc. Act.

### 7.3.3  Outline Process Description of Meissner Plant

The Meissner plant at Hickson and Welch produced three isomers of mononitrotoluene (MNT) from the nitration reaction of toluene using a mixture of nitric and sulphuric acid. Byproducts of the reaction include dinitrotoluene and nitrocresols. In the original design, the nitrocresols were removed in a caustic wash stage, although this part of the process was revised in 1988 (See section 7.3.4.1).

Next, a series of continuous distillation stages separated the ortha-, meta-, and para-nitrotoluene (oNT, mNT and pNT) isomers, as shown in the drawing in Figure 7.8 (HSE 1994). The final crystallization stage extracted pNT leaving a liquor that was called "40% Whizzer" that was stored in tanks. Since the process was revised in 1988, this material then underwent batch distillation, to recover further MNT by boiling it off under vacuum.

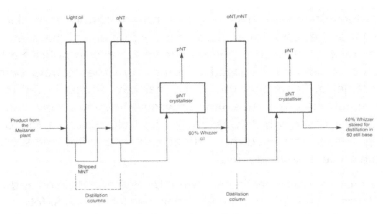

**Figure 7.8  Schematic Drawing of the Separation Stages**

The batch distillation took place in a horizontal steel drum called "60 Still Base," as shown in the schematic drawing in

(HSE UK 1994). This 45 m³ vessel was 7.9m (26 ft) long, 2.7m (9 ft) diameter, and was fitted with three steam heater batteries (series of tubes) towards the bottom, with the lower battery about 230mm (9") above the tank bottom and the other two some 430mm (17") above the lower coil. The heater batteries were heated with steam that came from a 27.6 barg (400 psig) supply pressure, via a regulator that reduced the pressure below 6.9 barg (100 psig), with a relief valve set to lift at 100 psig.

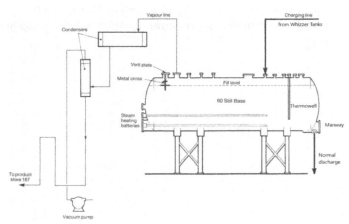

**Figure 7.9  Schematic Drawing of 60 Still Base**

The 60 Still Base was fitted with a thermocouple, mounted in a thermowell from the top of the vessel and operated under a vacuum when the distillation process was underway.

Normal procedure was for distillation to be carried out in two stages. Initially, the vessel was filled to depth of some 2.16m (85 inches) and operators took a sample for thermal analysis before the vacuum was started and heat applied. Typically, the distillation continued until there was a 50:50 mixture of the volatile MNTs to the less volatile DNT and nitrocresols. The thermal analysis was used to check for thermal stability of the residues that was used as a basis for evaluating how far the distillation would proceed.

Once this first stage was complete, the vacuum would be broken, and a second charge of Whizzer Oil added before additional thermal testing was done. If the results were acceptable, the vacuum would be applied and distillation continued until, again, there was a 50:50 mixture of volatiles to non-volatiles. The final residue in 60 Still Base was taken away in a tanker for disposal by off-site incineration.

### 7.3.4 History of Meissner Plant Prior to Incident

#### 7.3.4.1 Process

There were two Meissner plants at the Castleford facility. Meissner I dated from 1962 with a capacity of 20 tonnes per day and Meissner II was commissioned in 1972 with a capacity of 60 tonnes per day. The products from the nitration reaction were first washed with water, to remove the residual acids, and up until 1988, this was followed by a caustic wash stage to remove the byproduct nitrocresols, which was treated and then discharged as an effluent to the local river.

In 1988, the caustic wash stage was removed for environmental and safety reasons. This meant that the nitrocresols were carried further through the process. At the same time, they introduced the batch distillation of the 40 % whizzer oil using 60 Still Base.

The hazards associated with the potential thermal instability of substances such as DNT and nitrocresols were well known to the management at the site. Company specialists developed thermal stability tests that were reviewed in 1988 and 1989 when the process of batch distillation in 60 Still Base was introduced.

After the modification in 1988, problems started with residue accumulation in parts of the continuous distillation section of the process. These were stripped and cleaned out although operators also noted the accumulation of sludge in 60 Still Base, up to a depth of some 34cm (14 inches).

In June 1992, one of the senior process technologists expressed frustration about these problems and the impaired conditions on the continuous MNT distillation section of the process. A memo was written stating: "It is my view that we are within five years of a major accident on the MNT distillation system."

At some point (date unknown), the steam regulator on the supply to the heater battery became non-operational. This was overcome by someone opening a bypass valve around the regulator until the relief valve started to lift.

On 10 September 1992, some 11 days before the incident, two of the 40% whizzer oil tanks were fully emptied to the 60 Still Base, in order to clean them out for a change of product. Being a vacuum vessel, this allowed the residual sludge from these tanks to be sucked into 60 Still Base, further increasing the level of solids in the vessel. On Thursday, 17 September, the removal of the sludge from 60 Still Base was discussed by a shift manager and area manager. Since this vessel had never been cleaned out before, they discussed some of the practical measures required including removal of some steps and provision of a skip (steel dumpster) to collect the sludge.

A batch distillation of the material in the 60 Still Base then took place on Saturday 19 September in preparation for the work to start the cleanout of 60 Still Base the following Monday, 21 September 1992.

### 7.3.4.2   *Organization*

In parallel with the process changes, several organizational changes took place at the Castleford site. The original management structure was a traditional, hierarchical one, where plant managers managed individual plants and each shift comprised a shift supervisor and a small number of shift operators.

This was changed in August 1992 to a "matrix" system where the plant manager role was eliminated and the senior operators were appointed as team leaders, reporting to an area manager. Staff had to apply for their new job roles and the new system was not fully understood. Further details of these changes are provided in an IChemE Loss Prevention Bulletin article, *Failure to manage organisational change – a personal perspective* (Lynch/IChemE 2019).

### 7.3.5 The Incident

**PREPARATION**

The site shift manager and area manager discussed the removal of residue from 60 Still Base on Thursday, 17 September 1992. Over the following weekend, the contents were distilled to reduce the level of volatile MNTs in the usual way. This was completed and the still base was allowed to cool before the remaining liquid material was pumped out to storage.

At 09:45 on Monday 21 September, the area manager instructed staff to apply steam to the heater batteries, in order to soften the sludge, and advised that the temperature was not to exceed 90 °C. By 10:15, a skip had been obtained to hold the waste sludge, and a scaffold was erected beneath the manhole cover.

A permit to work was issued by the team leader, initially to a fitter, although he then left for an early lunch. A second permit was then provided for the operators to remove the manhole cover, which took about 30 minutes. Some of the material was scooped out of the vessel and found to be gritty, with the consistency of "soft butter". It is understood that the material was not tested further, and the area manager assumed it was a thermally stable tar.

**RAKING**

Two process technicians then started to remove material using a 2.5m long metal rake that was "found" on the ground nearby. After about an hour, (at approximately 12:50) a 2m length of sludge had been removed and the team decided to make an extension for the rake so it could reach further back into the still base. At the same time, the fitter returned from lunch and noted that the inlet to the still base had not been sealed off.

The team leader then provided the fitter with a permit to perform this task by blanking the inlet line.

By this time, the temperature in 60 Still Base was showing 48 °C on the control room panel and the area manager instructed the steam to be isolated on the heater battery, which was shut off at about 13:00.

## IGNITION

One technician was left on the scaffold and continued raking until about 13:20, when he noted a blue light followed immediately by an orange flame. He leapt off the scaffold and saw a horizontal, incandescent jet of flame that was projected towards the Meissner plant control building and the office, accompanied by a loud roar (heard both on and off site). The fitter, who was on the other side of the vessel fitting the inlet blank, also saw a vertical jet of flame emanating from the top vent of the vessel.

## FIRE!

A jet of flame emanated both horizontally from the manway and vertically from the vessel vent. The jet only lasted for some 25 seconds but it largely destroyed the control room, caused four fatalities, and set the main office block ablaze, causing an additional fatality in women's washroom.

The report (HSE UK 1994) continues to describe the emergency response, damage, and injuries.

### 7.3.6    Immediate Causes

Ignition of the contents of 60 Still Base was due to exothermic decomposition of thermally unstable residues (dinitrotoluene/ nitrocresols) in the sludge. This ignited the MNT vapors in the drum, and it was estimated that the flame, which lasted for some 25 seconds, had a temperature as high at 2,300 °C some 6 m (20 feet) from the manway.

Several factors contributed to the immediate cause of this thermal runaway:

- No thermal stability testing of the material in 60 Still Base had been conducted, even though management knew that unstable material was likely to be present. The onset temperature (the critical temperature at which thermal runaway is initiated) was therefore unknown.

- The thermocouple with the 60 Still Base was not long enough to reach the sludge and was therefore only measuring the temperature of the vapor space. The temperature of the sludge was likely to have been much hotter.

- The temperature of the condensing steam in the heater battery should have been limited to about 160 °C since the pressure was regulated to below 6.9 barg (100 psig). However, the regulator was broken, and the bypass valve had been opened. The steam pressure relief valve should have opened at 6.9 barg (100 psig), but this was subsequently found to operate at 9.0 bar (135 psig), at which the steam condensing temperature would have been 175 °C.

## 7.3.7    Lessons Learned Relevant to Abnormal Situation Management

While this event occurred during a period of maintenance, many of the underlying and root causes, and associated lessons learned relate to the management of abnormal situations as discussed throughout this book. The report contains full details (HSE UK 1994).

### 7.3.7.1    Abnormal Situation Recognition

The 60 Still Base had never been cleaned out in its 30 years and it must have been clear to the site personnel that the process of cleaning it out was an abnormal situation. As such, the entire operation should have been subjected to a full and detailed risk assessment, but it was not (see 7.3.7.5). Although the issue of thermal stability was known about within the company, it is unlikely that the people authorizing and conducting the work had the same level of understanding, particularly since no permit was issued for raking out the material.

### 7.3.7.2    Design and Change Management

The original design of the equipment included a separation stage whereby potentially thermally unstable nitrocresols were removed at an earlier stage of the process. This was modified in 1988 such that this material was carried through the process and ended up in 60 Still Base. Although this was recognized as a potential problem at the time, plant trials indicated that the thermal stability of the substances was greater than had been predicted by the simulations. However this incident demonstrated that the thermal stability was not actually greater.

The design of the 60 Still Base included a thermocouple that, under normal circumstances would measure the temperature of the liquid in the drum. However, it was positioned above the heater batteries and under the situation where the liquids had been removed from the drum; it was in the gas space and did not measure the sludge temperature. This meant that the sludge could have been at a much higher temperature than the specified 90 °C, and the surface temperature of the steam batteries could have been higher still. The failure of the steam regulator and operating it on bypass should have been the subject of a management of change (MOC) procedure.

The effect of the change in design (removal of the early stage to separate nitrocresols) and change in procedures (operating steam regulator on bypass) should have been reviewed in the light of the operating experience of the unit. The changes should also have been a factor in the procedure that was adopted for the removal of the residues. Principles of Inherently Safe Design would encourage the removal of potentially unstable materials as early as possible in the process, or ideally to prevent their formation in the first place.

### 7.3.7.3 *Management of Organizational Change (MOOC)*

There had been several recent changes to the organization and the Castleford site, and the reporting lines had changed significantly. The report states that the newly designated team leaders had not received adequate training and that there were a number of errors on the permit and preparation of the safe system of work. The newly appointed area manager did not check the system of work, or the permits, and his attention was distracted due to other problems.

MOOC should consider the level of labor and skills available to safely deal with abnormal situations, particularly when switching from a hierarchical to a matrix organization.

### 7.3.7.4 *Human Factors and Culture*

Employees were pressured to use the 60 Still Base quickly in order to process the high stocks of whizzer oil. It may be that there was pressure to clear out the drum quickly, so that the impact on production would be limited.

Pressure to achieve high levels of production must not compromise the normal processes of risk assessments, particularly when abnormal conditions are encountered.

### 7.3.7.5 Procedures

Risk assessment and safe system of work should be employed for all activities to ensure that they are conducted with an acceptable degree of risk. Procedures to decontaminate and clean out residues from tanks and vessels require special attention due to the high potential for incidents. A separate assessment of the risks is required to ensure work can be conducted safely in addition to issuing permits to allow the mechanical work to be done.

Prior to this incident, possibly due to a lack of planning or understanding, the site did not conduct a detailed assessment of the risks associated with cleaning out the sludge. This was due, in part, to some of the recent organizational changes (See 7.3.7.3). Key factors included a lack of thermal stability testing; decision to heat up the sludge; failure to test for flammable atmosphere in 60 Still Base; and incomplete isolation of the 60 Still Base.

The steam pressure relief valve was found to be operating above its set point and its lifting point was being used as a measure to set the pressure regulator. Pressure relief valves must not be used as a control device. A HAZOP or other risk assessment should have identified that these devices were part of a critical safety system due to the potential thermal instability of the sludge in the 60 Still Base. As such, they should both have been on a regular test schedule.

Abnormal situations can lead to pressure on individuals to forgo normal procedure; however, it is during these occasions that procedures should be followed without deviating.

### 7.3.7.6 Process Monitoring and Control/ Instrument Failures

The issue of the thermocouple not being below the liquid surface (in this case the sludge) and therefore not accurately measuring the temperature is a known contributory cause of several incidents.

### 7.3.7.7 Training /Knowledge and Skill

The proper level of understanding of the thermal stability issues were not reflected by the decision and method used to clear out the material from 60 Still Base. Stability tests on the material were not conducted and the decision to heat up the sludge was a key factor that led to the incident.

It is important that people with the appropriate knowledge and skill, including subject matter experts, be consulted when encountering abnormal situations.

### 7.3.8    Epilogue

Following the incident, investigation, and clean up, Hickson and Welch modified the process such that the batch distillation of the whizzer oil was eliminated. The company devised a system involving distillation and fractional crystallization and installed it before restarting operations.

# APPENDIX A
# MANAGING ABNORMAL
# SITUATIONS –TRAINING MATERIALS

## Online Training Modules

As part of the development of this book, five online training modules were developed relating to abnormal situations. These training modules can be used by supervisors, plant engineers and trainers to help train operating teams in the diagnosis of an abnormal situation. The modules allow the trainer to step through specific abnormal situations and discuss diagnosis, actions to take, learning and relevance to their operation with the team members who are being trained. The modules are in Microsoft PowerPoint® format.

## Access Information

To access this online material, go to:

*www.aiche.org/ccps/publications/Situations*

Enter password:

*ASM2022*

# APPENDIX B
# ASM JOINT RESEARCH AND DEVELOPMENT CONSORTIUM: BACKGROUND

The Abnormal Situation Management® Consortium (ASMC) is a group of leading process industry companies and universities that have jointly invested $50 million in research and development to create knowledge, tools, and products designed to prevent, detect, and mitigate abnormal situations.

Abnormal situations are undesired plant disturbances with which the control system is not able to cope. Consequently, a human operator must intervene to supplement the actions of the control system. An abnormal situation could be quickly rectified by operator action, or it may escalate to a critical incident in which equipment damage, serious injury, fatality, and/or significant impact to the community may result. Most abnormal situations do not escalate to become critical incidents, but even minor excursions can cause production losses, quality downgrades, and higher production costs.

While "human error" is often blamed, the ASM Consortium has shown that incidents are caused by multiple factors: but preventative solutions are available based on human factor design of the systems and operator environment.

The Consortium began as a joint program of the U.S. National Institute of Science and Technology and the Process Industry. It based its research priorities after categorizing the failure modes from 42 detailed public and private incident reports and many visits to plant sites.

The rigor of the Consortium's research is considerable. University researchers with human factors expertise conduct studies to determine likely solutions to issues presented. If successful, prototype solutions are tested in member plants, and improvements are verified. A team of 17 Industry Members, 7 Universities, and 6 Human Factors Associate members contributed to its archive of over 500 technical reports and more than 500 case histories, presentations, and in-kind member contributions, available to ASM members.

The Consortium has published over 35 papers and 5 guidelines.

*ASM and Abnormal Situation Management Consortium are U.S. registered Trademarks of Honeywell International, Inc.*

For more information, see the Consortium's web site at http://www.asmconsortium.net

# REFERENCES

Note: These references and associated internet websites (if applicable) were current at the time they were accessed during this guideline's preparation (2020-2021).

1.  Abnormal Situation Management Consortium®. ASM® Consortium, available at: https://www.asmconsortium.net/defined/definition/Pages/default.aspx
2.  Airfrance Website. *Airbus Service Bulletins – Pitot Tubes*, accessed June 2021, https://corporate.airfrance.com/en/pitot-probes
3.  American National Standard Institute/International Society of Automation 2009. ANSI/ISA-18.2 2009 - *Management of Alarm Systems for the Process Industries*, revised 2016.
4.  American Petroleum Institute 2010. *API 1167 2010 - Pipeline SCADA Alarm Management,* revised 2016.
5.  American Petroleum Institute 2014. *API RP 585 - Pressure Equipment Integrity Incident Investigation*, Recommended Practice 585. Washington D.C.
6.  American Petroleum Institute 2016. *ANSI/API RP 754 - Process Safety Performance Indicators for the Refining and Petrochemical Industries*, Recommended Practice 754, 2nd Edition, Washington D.C.
7.  Baker Engineering and Risk Consultants, Inc. 2021. *Schematic Illustration of the Relationship between Normal and Emergency Operating Limits, and Emergency Shutdown,* BakerRisk.
8.  Bedny & Meister 1999. *Theory of Activity And Situation Awareness,* Int J Cognitive Ergonomics 3 (1) 63-72, reference from Stanton NA.
9.  Bransby ML & Jenkinson J 1998. 'The management of alarm systems,' Health and Safety Executive. HSE UK Contract Research Report 166/1998, available at: https://www.hse.gov.uk/research/crr_pdf/1998/crr98166.pdf
10. Broadribb MP 2003. *Lessons Learned from Augusta: A Case History*, Proceedings of the Center for Chemical Process Safety 18th Annual International Conference, Scottsdale, AZ, 2003.
11. Broadribb MP et al 2009. *Cheddar or Swiss? How Strong are your Barriers? (One Company's Experience with Process Safety Metrics)*, Proceedings of the Global Congress on Process Safety, April 2009.

12. Bullemer, PT & Laberge, JC 2009. *Common Operations Failure Modes in the Process Industries*, 12th Annual Symposium, Mary Kay O'Connor Process Safety Center, "Beyond Regulatory Compliance: Making Safety Second Nature", available at: http://pscmembers.tamu.edu/wp-content/uploads/Bullemer.pdf

13. Bullemer PT, Kiff L and Tharanathan A 2010a. "Common Procedural Execution Failure Modes During Abnormal Operations", Mary Kay O'Connor Process Safety Center International Symposium, College Station, TX, 2010.

14. Bullemer PT, Kiff L & Tharanathan A 2010b. *Common procedural execution failure modes during abnormal situations,* Mary Kay O'Connor Process Safety Center Symposium, Texas, USA. Accessed 11 June 2021, available at: http://pscmembers.tamu.edu/wp-content/uploads/009_Bullemer.pdf

15. Bullemer P, Hajdukiewicz J & Burns C 2010. *Effective Procedural Practices.* Abnormal Situation Management® Consortium. Minneapolis: CreateSpace Independent Publishing Platform, ISBN 978-1452893877.

16. Bullemer PT & Reising D 2015. *Effective Console Operator HMI Design* (2nd Edition). Abnormal Situation Management® Consortium. Minneapolis: CreateSpace Independent Publishing Platform, ISBN 978-1514203859.

17. Bullemer PT 2020. *Effective Operations Practices.* Abnormal Situation Management® Consortium. Houston: ISBN 978-1703706512.

18. Bullemer PT & Reising D 2021. *Effective Change Management Practices in HMI Development.* Abnormal Situation Management® Consortium. ISBN 979-8567792957.

19. Bureau for Analysis of Industrial Risks and Pollutions (BARPI) database. Research and Information on Accidents (ARIA) database, an Analysis—a searchable database of incidents and other reference material. Accessed June 2021, https://www.aria.developpement-durable.gouv.fr/?lang=en

20. Bureau of Enquiry and Analysis for Civil Aviation Safety (BEA). *Final Report on the accident on 1st June 2009 to the Airbus A330-203 registered F-GZCP operated by Air France flight AF 447 Rio de Janeiro – Paris.* Accessed June 2021, https://www.bea.aero/docspa/2009/f-cp090601.en/pdf/f-cp090601.en.pdf

21. Center for Chemical Process Safety. *Process Safety Beacon.* American Institute of Chemical Engineers (AIChE), CCPS, accessed February 2021, available at: https://www.aiche.org/ccps/process-safety-beacon

22. Center for Chemical Process Safety 2021. *Process Safety Beacon.* CCPS, accessed June 2021. https://www.aiche.org/ccps/process-safety-beacon

23. Center for Chemical Process Safety. *Process Safety Glossary*, available at: https://www.aiche.org/ccps/resources/glossary/process-safety-glossary/abnormal-situation

24. Center for Chemical Process Safety. *PSID: Process Safety Incident Database.* AIChE,CCPS, accessed February 2021, available at: https://www.aiche.org/ccps/resources/psid-process-safety-incident-database

25. Center for Chemical Process Safety. *PSID* 2018-2: *Process Safety Incident Database*. Produced by CCPS, accessed June 2021, available at: https://www.aiche.org/ccps/resources/psid-process-safety-incident-database
26. Center for Chemical Process Safety 1995a. *Guidelines for Safe Process Operations and Maintenance*, AIChE, CCPS, New York.
27. Center for Chemical Process Safety 1995b. *Guidelines for Safe Process Operations and Maintenance*, 1995; p 113, citing "Large Property Damage Losses in the Hydrocarbon-Chemical Industries", Marsh McLennan, 14th edition, CCPS, New York 1992.
28. Center for Chemical Process Safety 1996. *Guidelines for Writing Effective Operating and Maintenance Procedures*, AIChE, CCPS, New York.
29. Center for Chemical Process Safety 1999. *Guidelines for Chemical Process Quantitative Risk Analysis*, 2nd edition, AIChE, CCPS, New York.
30. Center for Chemical Process Safety 2001. *Layer of Protection Analysis-Simplified Process Risk Assessment*.
31. Center for Chemical Process Safety 2004. *Guidelines for Preventing Human Error in Process Safety*. AIChE, CCPS, New York.
32. Center for Chemical Process Safety 2006. *Human Factors Methods for Improving Performance in the Process Industries*, AIChE, CCPS, New York.
33. Center for Chemical Process Safety 2007a. *Guidelines for Risk Based Process Safety*. AIChE, CCPS, New York.
34. Center for Chemical Process Safety 2007b. *Guidelines for Safe and Reliable Instrumented Protective Systems*, AIChE, CCPS, New York.
35. Center for Chemical Process Safety 2007c. *Human Factors Methods for Improving Performance in the Process Industry*.
36. Center for Chemical Process Safety 2008a. Concept book, *Incidents That Define Process Safety*.
37. Center for Chemical Process Safety 2008b. *Guidelines for Hazard Evaluation Procedures*. 3rd edition, AIChE, CCPS, New York.
38. Center for Chemical Process Safety 2008c. *Guidelines for Management of Change for Process Safety*.
39. Center for Chemical Process Safety 2008d. "Three Mile Island Nuclear Reactor Core Meltdown March 28, 1979," *Incidents that Define Process Safety*.
40. Center for Chemical Process Safety 2009. *Guidelines for Process Safety Metrics,* AIChE, CCPS, New York.
41. Center for Chemical Process Safety 2010. *A Practical Approach to Hazard Identification for Operations and Maintenance Workers*, AIChE, CCPS, New York.
42. Center for Chemical Process Safety 2011a. *Recognizing Catastrophic Incident Warning Signs in the Process Industries*, AIChE, CCPS, New York.
43. Center for Chemical Process Safety 2011b. *Conduct of Operations and Operational Discipline: For Improving Process Safety in Industry*, AIChE,CCPS, New York.

44. Center for Chemical Process Safety 2013. *Guidelines for Managing Process Safety Risks During Organizational Change.*
45. Center for Chemical Process Safety 2015. *Guidelines for Defining Process Safety Competency Requirements,* AIChE, CCPS, New York.
46. Center for Chemical Process Safety 2016c. *Guidelines for Integrating Management Systems and Metrics to Improve Process Safety Performance,* AIChE, CCPS, New York.
47. Center for Chemical Process Safety 2017a. *Guidelines for Asset Integrity Management,* AIChE, CCPS, New York.
48. Center for Chemical Process Safety 2017b. *Guidelines for Safe and Reliable Instrumented Protective Systems.* AIChE, CCPS, New York.
49. Center for Chemical Process Safety 2017c. *Guidelines for Safe Automation of Chemical Processes*, 2nd Edition, New York.
50. Center for Chemical Process Safety 2018a. *Bow Ties in Risk Management: A Concept Book for Process Safety.*
51. Center for Chemical Process Safety 2018b. *The Business Case for Process Safety*, 4th edition, available at: https://www.aiche.org/sites/default/files/docs/embedded-pdf/3433_18_ccps_buscaseprocesssafety_0.pdf
52. Center for Chemical Process Safety 2018c. *Essential Practices for Creating, Strengthening, and Sustaining Process Safety Culture*, AIChE, CCPS, New York.
53. Center for Chemical Process Safety 2018d. *Process Safety Leading and Lagging Metrics: You Don't Improve What You Don't Measure*, AIChE, CCPS, New York.
54. Center for Chemical Process Safety 2018e. *Process Safety Metrics: Guide for Selecting Leading and Lagging Indicators*, Version 3.2, AIChE, CCPS, New York.
55. Center for Chemical Process Safety 2018f. *Recognizing and Responding to Normalization of Deviance*, AIChE, CCPS, New York.
56. Center for Chemical Process Safety 2019. *Guidelines for Investigating Process Safety Incidents*, 3rd Edition, AIChE,CCPS, New York.
57. Chemical Engineering January 2016. *Training Simulators Make Real Difference.*
58. Chemical Safety and Hazard Investigation Board. Reports and videos on major incidents, US CSB, accessed June 2021, https://www.csb.gov/investigations/completed-investigations/
59. Chemical Safety and Hazard Investigation Board 2005. "Refinery Explosion and Fire, BP Texas CityUS CSB Report, available at: https://www.csb.gov/file.aspx?DocumentId=5596
60. Chemical Safety and Hazard Investigation Board 2007. *Investigation Report, Refinery Explosion and Fire, March 23, 2005, BP Texas City*, US CSB, Washington D.C.

61. Chemical Safety and Hazard Investigation Board 2011. *Pesticide Chemical Runaway Reaction Pressure Vessel Explosion*, Investigation Report No. 2008-08-I-WV, US CSB, Washington D.C., January 2011.

62. Chemical Safety and Hazard Investigation Board 2018. *Safety Digest: CSB Investigations of Incidents During Startups and Shutdowns*, US CSB, Washington D.C.

63. Chemical Safety and Hazard Investigation Board 2019. *Gas Well Blowout and Fire at Pryor Trust Well 1H-9,* US CSB, Washington D.C.

64. Chemical Safety and Hazard Investigation Board 2021. *Safety Digest,* US CSB, accessed June 2021. https://www.csb.gov/news/

65. Civil Aviation Authority 2015. CAA CAP 795: *Safety Management Systems (SMS) Guidance to Organisations*, 2 February 2015, accessed June 2021. https://publicapps.caa.co.uk/modalapplication.aspx?appid=11&mode=det ail&id=6616

66. Crichton MT, et al 2000. "Training Decision Makers -Tactical Decision Games", *Journal of Contingences and Crisis Management,* Vol 8 No 4 Dec 2000.

67. Department of Transportation (DOT) Federal Aviation Administration (FAA) Advisory Circular. US DOT/FAA Advisory Circular: *Safety Management Systems for Aviation Service Providers*, accessed June 2021. https://www.faa.gov/documentLibrary/media/Advisory_Circular/AC_120-92B.pdf

68. Downes T 2017. *Past and Future of Abnormal Situation Management*, 62nd Annual Safety in Ammonia Plants and Related Facilities Symposium 2017.

69. Duguid, I. M. 1998a. *Analysis of Past Incidents in the Oil, Chemical and Petrochemical Industries*, Loss Prevention Bulletin, No. 142, p. 3, Institution of Chemical Engineers (IChemE), Rugby, UK.

70. Duguid, I. M. 1998b. *Analysis of Past Incidents in the Oil, Chemical and Petrochemical Industries*, Loss Prevention Bulletin, No. 143, p. 3, IChemE, Rugby, UK.

71. Duguid, I. M. 1998c. *Analysis of Past Incidents in the Oil, Chemical and Petrochemical Industries*, Loss Prevention Bulletin, No. 144, p. 26, IChemE, Rugby, UK.

72. Electronic Code of Federal Regulations (eCFR). Title 14: Aeronautics and Space. An electronic version of general and permanent rules published in the Federal Register, accessed June 2021. https://ecfr.federalregister.gov/current/title-14/chapter-I/subchapter-G/part-119/subpart-A

73. Embrey DE, Kirwan B, Rea K, Humphreys P and Rosa EA 1984. "SLIM-MAUD, An Approach to Assessing Human Error Probabilities Using Structured Expert Judgment, Vols I and II, NUREG/CR-3518." US Regulatory Commission, Washington.

74. Emerson White Paper 2019. *Alarm Rationalization,* October 2019, available at: https://www.emerson.com/documents/automation/white-paper-alarm-rationalization-deltav-en-56654.pdf

75. Energy Institute 2014. *Guidance on Crew Resource Management (CRM) and Non-Technical Skills Training Programmes.*

76. Engineers Equipment and Materials User Association 1999. EEMUA 191 1999: "Alarm Systems- A Guide to Design, Management and Procurement", revised 2013.

77. Engineers Equipment and Materials User Association 2002. *Ergonomic Design Guidance,* EEMUA.

78. Errington J, Reising DVC, Bullemer P, DeMaere T, Coppard D, Doe K & Bloom C 2005. "Establishing Human Performance Improvements and Economic Benefit for a Human-Centered Operator Interface: An Industrial Evaluation", Human Factors and Ergonomics Society 49th Annual Meeting, Orlando, FL, 2005.

79. Ethiopian Ministry of Transport Aircraft Accident Investigation Bureau 2019. Federal Democratic Republic of Ethiopia, *Aircraft Accident Investigation Preliminary Report Ethiopian Airlines Group B737-8 (MAX) Registered ET-AVJ.* Accessed June 2021. Available from: https://reports.aviation-safety.net/2019/20190310-0_B38M_ET-AVJ_PRELIM.pdf

80. European Agency for Safety and Health at Work (EU-OSHA) 2013. *Diverse cultures at work: ensuring safety and health through leadership and participation.*

81. European Commission Major Accident Reporting System (eMARS) database. A searchable database of incidents in the EU. Accessed June 2021. https://emars.jrc.ec.europa.eu/en/emars/accident/search

82. European Process Safety Centre *Learning Sheet.* Produced by the EPSC, accessed June 2021, http://epsc.be/Learning+Sheets.html

83. European Union Aviation Safety Agency. *EASA: Airworthiness Directive AD No.: 2009-0195, Navigation – Airspeed Pitot Probes – Replacement.* Accessed June 2021, available at: https://ad.easa.europa.eu/blob/easa_ad_2009_0195.pdf/AD_2009-0195_1

84. Flin et al 2003. Development of the NOTECHS (Non-Technical Skills) system for assessing pilots' CRM skills. *Human Factors and Aerospace Safety,* 3, 95-117.

85. Furlong Andy, Press Release 2014. Institution of Chemical Engineers *Remember Bhopal,* IChemE, available at: https://www.icheme.org/about-us/press-releases/remember-bhopal/

86. Global Congress on Process Safety 2017. *Closing the Holes in the Swiss Cheese Model – Maximizing the Reliability of Operator Response to Alarms,* exida consulting DuPont Protection Solutions and Beville Engineering, Inc., GCPS 2017 Conference.

87. Gupta JP 2004. *Bhopal and the Global Movement on Process Safety.* Institution of Chemical Engineers (IChemE) Symposium Series 150 2004.
88. Hailwood M 2020. *Release of hazardous vapours at LG Polymers chemical plant in Visakhapatnam, Andhra Pradesh, India,* Loss Prevention Bulletin, IChemE, August 2020.
89. Health and Safety Executive 1994. *A report on the investigation by the Health and Safety Executive into the fatal fire at Hickson & Welch Ltd., Castleford on 21 September 1992,* HSE Books, ISBN 0-7176-0702-X, 1994. Accessed June 2021, available at: https://www.icheme.org/media/13704/the-fire-at-hickson-and-welch-ltd.pdf
90. Health and Safety Executive 1997. *The Explosion and Fires at the Texaco Refinery, Milford Haven, 24 July 1994,* HSE Books, Her Majesty's Stationery Office, Norwich, UK, 1997.
91. Health and Safety Executive 1997. *The explosion and fires at the Texaco Refinery, Milford Haven, 24 July 1994,* HSE UK, ISBN 0 7176 1413 1, 1997
92. Health and Safety Executive website. "Case studies," HSE UK, accessed June 2021, https://www.hse.gov.uk/comah/sragtech/casestudyind.htm
93. Hemsley K & Fisher RE 2018. *History of Cyber Incidents and Threats to Industrial Control Systems,* Idaho National Laboratory, accessed February 2021, available at: https://www.osti.gov/servlets/purl/1505628
94. Institution of Chemical Engineers/Final Report 2008. The Buncefield Incident 11 December 2005, "The final report of the Major Incident Investigation," Board Volume 1, 2008. ISBN 978 0 7176 6270 8. IChemE, accessed February 2021, available from: https://www.icheme.org/media/13707/buncefield-miib-final-report-volume-1.pdf
95. Institution of Chemical Engineers 2021a. *Loss Prevention Bulletin.* Produced by the IChemE in the UK, accessed June 2021. http://www.icheme.org/lpb
96. Institution of Chemical Engineers 2021b. *Safety Lore.* Produced by the IChemE Safety Centre in the UK, accessed June 2021, https://www.icheme.org/knowledge/safety-centre/safety-lore/
97. International Electrotechnical Commission 2014. IEC 62682 2014. *Management of alarm systems for the process industries.*
98. International Association of Oil and Gas Producers 2018. *Process Safety – Recommended Practice on Key Performance Indicators,* IOGP Report 456, 2nd edition, London, UK.
99. International Association of Oil and Gas Producers 2020. *IOGP Report 501: Crew Resource Management for Well Operations Teams.*
100. International Organization for Standardization ISO 11064. ISO, 2000a; 2000b; 2002; 2004a; 2004b; 2005; 2006.
101. Jarvis R & Goddard A 2020. Webinar: *An Analysis of Common Causes of Major Losses in the Onshore Oil, Gas & Petrochemical Industries: Implications for Insurance Risk Engineering Surveys.* Oil Petrochemical and Energy Risks Association (OPERA), London, UK, June 2020.

102. Kletz T. "The ICI [Imperial Chemical Industries] Safety Newsletters," mainly issued by Trevor Kletz, accessed June 2021, https://www.icheme.org/membership/communities/special-interest-groups/safety-and-loss-prevention/resources/ici-newsletters

103. Kletz TA. Mary Kay O'Connor Process Safety Center. Available at: http://psc.tamu.edu/trevor-kletz

104. KNKT [i.e., National Transportation Safety Committee/NTSC, Indonesian, (per Wikipedia)]. *Preliminary Aircraft Accident Investigation Report* KNKT.21.01.01.04, PT. Sriwijaya Air Boeing 737-500; PK-CLC, Near Kepulauan Seribu, Jakarta, Republic of Indonesia, 9 January 2021. Accessed June 2021. http://knkt.dephub.go.id/knkt/ntsc_aviation/baru/pre/2021/PK-CLC%20Preliminary%20Report.pdf

105. Lynch M 2019. *Failure to manage organisational change – a personal perspective*, IChemE Loss Prevention Bulletin 267, June 2019.

106. Mosier KL et al 2016. Electronic Checklists: Implications for Decision Making, *Proceedings of the Human Factors Society 36th Annual Meeting, Atlanta, Georgia, 12th-16th October 1992.*

107. NAMUR NA 102 2003. *Alarm Management*, Issued by the User Association of Process Control Technology in Chemical and Pharmaceutical Industries (NAMUR).

108. National Aeronautics and Space Administration. NASA Control Room – Engine Research Building, available at: https://archive.org/details/C-1948-21873

109. National Fire Protection Association 2021. *Guide for Fire and Explosion Investigations*, NFPA 921, Quincy, MA.

110. National Transportation Safety Board Report. *Flight 173 DC-8 Crash in Portland, NTSB Investigation.*

111. Occupational Health and Safety Administration. OSHA 1910.119. *Process Safety Management of Highly Hazardous Materials Standard* (element l). OSHA US.

112. Occupational Health and Safety Administration. OSHA PSM Regulations 29CFR 1910.119. *Process safety management of highly hazardous chemicals*, OSHA US. https://www.osha.gov/laws-regs/regulations/standardnumber/1910/1910.119

113. Ostrowski SW & Hertoghe N 2019. *A HAZOP Methodology for Transient Operations*, accessed 25 June 2021, available at: https://epsc.be/Events/Past+Events/EPSC+Conference+2019_+Program-p-1705/_/08%20Nicolas%20Hertoghe,%20Exxon%20-%20HAZOP%20for%20Transient%20Operations.pdf

114. Ostrowski SW & Keim KK 2010. *Tame Your Transient Operations – Use a special method to identify and address potential hazards*, Chemical Processing, June 23, 2010. http://www.chemicalprocessing.com/articles/2010/123/

115. Plastics Europe 2018. *Styrene Monomer: Safe Handling Guide*, July 2018. Available at: https://www.plasticseurope.org/download_file/force/2360/181

116. Rasmussen J 1982. *Human errors: A taxonomy for describing human malfunction in industrial installations*, Journal of Occupational Accidents, 1982, 4, 311-335.

117. Reason, James 1990. *Human Error*, Cambridge University Press.

118. Reising D & Bullemer P 2009. "An Introduction to the ASMGuidelines", 2009 ASM Webinar, accessed 25 June 2021, available at: https://www.asmconsortium.net/Documents/2009%20ASM%20Displays%20GL%20Webinar%20v014.pdf

119. Republic of Indonesia National Transport Safety Committee 2018. *Preliminary Air Accident Investigation Report, Lion Airlines Boeing 737 (MAX)*, 29 October 2018, accessed June 2021. https://reports.aviation-safety.net/2018/20181029-0_B38M_PK-LQP_PRELIMINARY.pdf

120. Rockwell White Paper 2017. *Alarm Rationalization and Implementation*, June 2017, available at: https://literature.rockwellautomation.com/idc/groups/literature/documents/wp/proces-wp015_-en-p.pdf

121. Stanton NA, Chambers PRG & Piggott J 2001. *Situational awareness and safety*. Safety Science 39 189-204, accessed 12 Feb 2021, available at: https://bura.brunel.ac.uk/bitstream/2438/1804/1/Situation_awareness_and_safety_Stanton_et_al,_pdf.\

122. Swain AD and Guttmann HE 1983. 'Handbook of Human Reliability Analysis with Emphasis on Nuclear Power Plant Applications, NUREG/CR-1278', US Regulatory Commission, Washington.

123. Syed M 2015. *Black Box Thinking*, John Murray (Publishers), London.

124. Transportation Safety Board of Canada Railway Investigation Report 2014. R13D0054, *Runaway and main-track derailment, Lac-Mégantic, Quebec, 06 July 2013*. Accessed June 2021. Available from: https://www.bst-tsb.gc.ca/eng/enquetes-investigations/rail/2013/r13d0054/r13d0054.html

125. User Centered Design Services, Inc. (UCDS). Available at: https://mycontrolroom.com/services/abnormal-situation-management/

126. Wincek JC and Haight JM 2007. "Realistic Human Error Rates for Process Hazard Analyses", *Process Safety Progress*, Vol. 26, No. 2, June 2007.

# INDEX

HIRA, 5, 10, 17, 40, 47, 50, 67, 86, 91, 113, 132, 133, 157, 159, 162
Hickson and Welch, 199, 200, 201, 210
human factors, 6, 11, 22, 23, 24, 33, 44, 55, 97, 98, 119, 127, 142, 166, 184, 195, 198, 214
Human Machine Interface, 23, 91, 104, 108, 110
HMI, 23, 29, 39, 47, 64, 91, 92, 93, 98, 104, 108, 110, 119, 140, 178, 180, 184, 193, 194, 198
Human Reliability Analysis, 55

**I**

Incident Investigation, 11, 49, 147, 155, 158
instrument air, 2, 31, 66, 180
intervention, 2, 3, 4, 9, 23, 39, 120

**L**

leadership, 6, 24, 44, 46, 50, 53, 64, 87, 89, 101, 126, 128, 145, 159, 162
loss of containment, 4, 5, 45, 104, 134

**M**

Management of Change, 36, 39, 41, 43, 49, 52, 57, 62, 91, 111, 129, 150, 153, 190, 196
Management of Organizational Change, 150, 153, 208
management review, 162, 163
Metrics, 7, 9, 49, 53, 103, 109, 125, 129, 130, 148, 156, 158, 163, 195
Milford Haven, 20, 184, 198
modification, 21, 62, 185, 189, 197, 204

**N**

near-misses, 7, 51, 106, 148, 157, 158, 182
Normalization of Deviance, 36, 44
nuisance alarms, 42, 43, 75, 117, 119

**O**

Oklahoma Well Blowout, 42, 76
operators, 7, 12, 17, 19, 22, 23, 25, 32, 39, 44, 46, 61, 64, 71, 74, 78, 87, 97, 103, 106, 120, 135, 138, 140, 151, 153, 195, 204
overfilled, 20, 37, 185, 192

**P**

Philadelphia Energy Solutions, 5
polymer, 15, 16, 17, 74
polystyrene, 77
power, 30, 31, 55, 61, 65, 66, 93, 114, 132, 184, 190, 194
Pressure Safety Valve, 70, 156, 158
Pre-Startup Safety Review, 129, 150, 154
Process Hazard Analysis
PHA, 74
Process Safety Incident Database, 51, 166
Process Safety Management, 151, 167
PSM, 30, 167
protection layers, 10, 11, 113, 119, 172

**R**

RAGAGEP, 33, 50, 86, 92
Reliability Centered Maintenance, 70
relief system, 18, 21, 35, 55, 57, 189
relief valve, 2, 16, 57, 82, 114, 120, 192, 202, 204, 207, 209
reputational damage, 5, 6, 11, 13, 34
Risk Based Process Safety, 22, 27, 47, 102, 113, 129, 134, 147, 150, 155, 158, 162, 167
runaway reaction, 18, 31, 59, 62, 69, 77, 79, 93, 113, 206
runaway train, 2

**S**

Safe Operating Limits, 10, 22, 25, 39, 40, 47, 116, 131
Safety Instrumented System, 120, 156, 158

# T

# W